Advancing Maths for AQA

MECHANICS 3

Ted Graham, Aidan Burrows and Joan Corbett

Series editors
Roger Williamson   Sam Boardman   Graham Eaton
Ted Graham   Keith Parramore

Heinemann Educational Publishers
a division of Heinemann Publishers (Oxford) Ltd,
Halley Court, Jordan Hill, Oxford OX2 8EJ

OXFORD JOHANNESBURG BLANTYRE MELBOURNE
AUCKLAND SINGAPORE GABORONE PORTSMOUTH NH (USA)
CHICAGO

First published in 2001

05 04 03 02 01
10 9 8 7 6 5 4 3 2 1

ISBN 0 435 51308 7

Cover design by Miller, Craig and Cocking

Typeset and illustrated by Tech-Set Limited, Gateshead, Tyne & Wear

Printed and bound by Scotprint in the UK

Acknowledgements
The publishers and authors acknowledge the work of the writers, Ray Atkin, John Berry, Sam Boardman, David Burghes, Derek Collins, Tim Cross, Ted Graham, Phil Rawlins, Tom Roper and Rob Summerson of the *AEB Mathematics for AS and A-Level* Series, from which some exercises and examples have been taken.

The publishers' and authors' thanks are due to the AQA for permission to reproduce questions from past examination papers.

The answers have been provided by the authors and are not the responsibility of the examining board.

About this book

This book is one in a series of textbooks designed to provide you with exceptional preparation for AQA's new Advanced GCE Specification B. The series authors are all senior members of the examining team and have prepared the textbooks specifically to support you in studying this course.

Finding your way around

The following are there to help you find your way around when you are studying and revising:

- **edge marks** (shown on the front page) – these help you to get to the right chapter quickly;
- **contents list** – this identifies the individual sections dealing with key syllabus concepts so that you can go straight to the areas that you are looking for;
- **index** – a number in bold type indicates where to find the main entry for that topic.

Key points

Key points are not only summarised at the end of each chapter but are also boxed and highlighted within the text like this:

For every action, there is an equal but opposite reaction

Exercises and exam questions

Worked examples and carefully graded questions familiarise you with the specification and bring you up to exam standard. Each book contains:

- Worked examples and Worked exam questions to show you how to tackle typical questions; Examiner's tips will also provide guidance;
- Graded exercises, gradually increasing in difficulty up to exam-level questions, which are marked by an [A];
- Test-yourself sections for each chapter so that you can check your understanding of the key aspects of that chapter and identify any sections that you should review;
- Answers to the questions are included at the end of the book.

Contents

CHAPTER 1

Mathematical modelling in mechanics

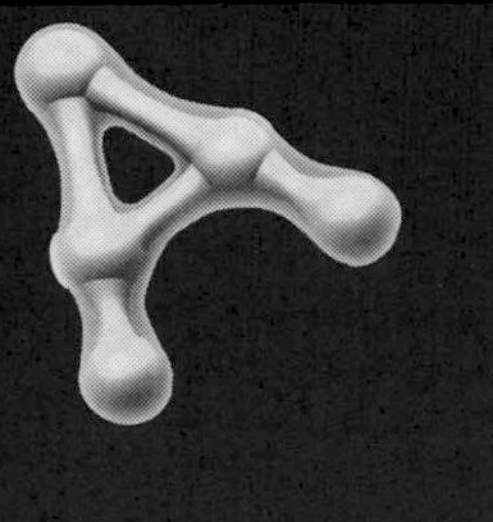

Learning objectives

After studying this chapter you should:

- be aware of the mathematical modelling cycle
- understand the types of assumptions used when modelling problems
- be aware of the difference between particle and rigid body models
- be familiar with some of the terminology used in mechanics.

1.1 Introducing modelling

Mathematical modelling describes the process of obtaining a solution to a real world problem. Many real problems can be very complex and so the idea of creating a mathematical model is to simplify the real situation, so that it can be described using equations or graphs. These equations or graphs are referred to as a mathematical model. These mathematical models can provide solutions to the original problem. It is often necessary to interpret these answers in the context of the original problem and to check that the answers that you have obtained are reasonable.

Examples of problems where modelling could be used:

- to determine the maximum speed of a car round a bend,
- to help define the design requirements of a sports stadium,
- to evaluate new design options for a mountain bike,
- to work out how to send a space station into orbit.

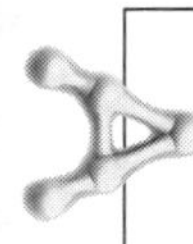

This whole process is often referred to as the mathematical modelling cycle because it may be necessary to repeat the process, creating better models of reality until a satisfactory solution is obtained.

The key stages of the mathematical modelling cycle are shown in the diagram on the next page.

The mathematical modelling cycle

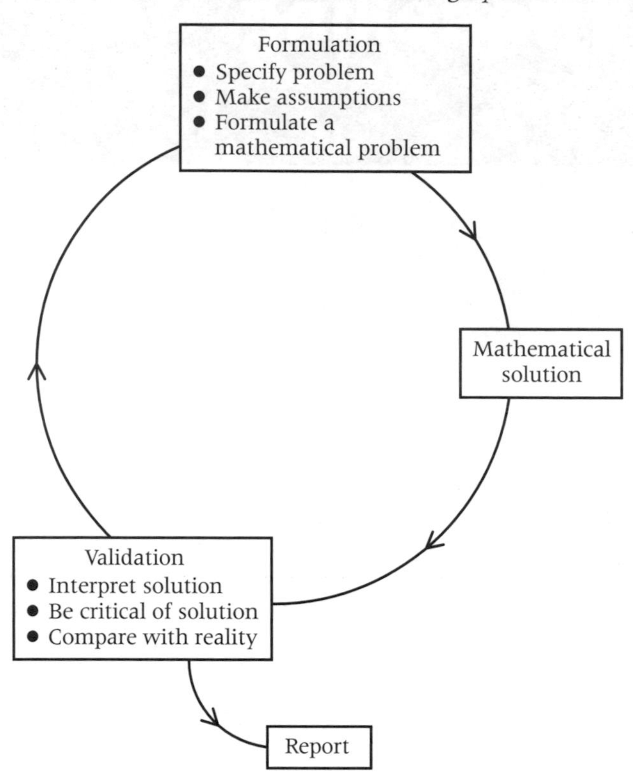

Each stage of this cycle is now considered in more detail.

1.2 The modelling cycle

Formulation

This stage of the modelling cycle consists of three distinct activities:

- specifying the problem,
- making assumptions,
- formulating a mathematic problem.

Specify problem

This is the starting point of the modelling cycle. Very often these problems will be very open-ended, unlike the types of problems that you will find in textbook exercises. Your first task is to make sure that you understand the problem and have decided what you need to find to solve the problem.

Make assumptions

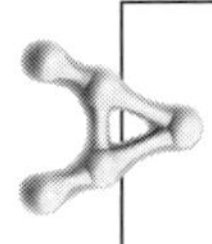

The next part of the formulation stage is to make assumptions to simplify the problem. Some of the assumptions that you will often make in mechanics are now considered, in the context of a body that remains at rest or moves in some way.

Assume that the body under consideration is a **particle**. This means that you assume that the body has no size, but does have a mass. So any rotation of the body is ignored, any forces will all act in the one place and the position is precisely defined. It may be quite reasonable to model a car as a particle if it is being driven a large distance or to model a ball as a particle.

A more sophisticated model is that of a **rigid body**. Here the body is assumed to have size, but to be rigid, so that it does not compress or change shape. The simplest rigid body is a rod. With a rigid body the forces will probably not all act at the same place and this must be taken into account. If you want to model the motion of a rolling ball it will be important to consider its size and shape, so a rigid sphere may be a suitable body.

A particle has no size but does have a mass.

A rigid body has size but does not change shape when forces are applied to it.

There are some factors (such as air resistance, lift on a golf ball, friction, resistance in a pulley) that may make a problem hard to describe and solve mathematically. We often ignore these factors in order to develop a simple mathematical model, but may have to incorporate them at a later stage if our results do not give good predictions. Further examples of things that we might ignore are the mass of a string, friction, that a string might be elastic and stretch.

Any assumptions made should be clearly recorded so that they can be reviewed later in the process.

Some keywords that you will find have particular meanings in a modelling context are listed below:

- smooth — no friction,
- rough — friction present,
- light — has no mass,
- inelastic — does not stretch,
- inextensible — does not stretch.

Formulate a mathematical problem

In this phase of the modelling cycle you turn the open-ended problem that you started with into a focused mathematical problem that you can solve using mathematical techniques.

You may also need to collect some data that is relevant to the problem that you have to solve.

You may also need to introduce algebraic variables to represent physical quantities that are important in the context of the problem. For example:

m the mass of the body,
u the initial velocity of the body,
etc.

Finally in the formulation stage there are a number of standard mathematical models to describe physical phenomena, for example to calculate the friction present, the size of the gravitational attraction, the force exerted by a stretched spring. You will learn about these models as you work through the mechanics modules and be able to include them in any modelling that you do.

Mathematical solution

This is where you obtain an actual solution, by carrying out calculations, solving equations and using other mathematical techniques. As you learn more pure mathematics you will be able to deal with more sophisticated mathematical models. When you have obtained a solution you move on to the interpretation and validation phase.

Validation

This is where you interpret the mathematical answers in the context of the original problem. This is where you will probably reach a conclusion of some kind. For example you may state that it is not safe for a lorry to drive over a bridge, that a car was breaking a speed limit or that it is impossible to hit a tennis ball over the net from a certain position.

It is also important to validate your answer, to check that it is reasonable. This may require you to make some observations of reality or to carry out an experiment to validate your predictions or it may simply be that the model you have created produced totally unrealistic solutions. For example if you calculate that a cyclist rolling down a hill will reach a speed of 70 mph after travelling 100 m, by ignoring air resistance, then it is clear that you need to go back to your original assumptions and try to include air resistance.

If you do need to go back to your original assumptions and reformulate your model, then you are embarking on a second cycle, which should lead to a more realistic solution. Finally when you obtain a satisfactory solution, you will need to report on your conclusion, describing how you reached this conclusion.

Worked example 1.1

How far apart should speed bumps be placed so that traffic does not reach a speed greater than 30 mph?

Solution

Assume that:

- a car is a particle,
- all cars slow down according to the table of stopping distances in the Highway Code,
- all cars speed up at the same rate as they slow down.

Gathering data from the Highway Code gives a figure of 23 m as the stopping distance at 30 mph.

Based on our assumptions we formulate the mathematical model:

$$\text{Distance between speed bumps} = 2 \times \text{stopping distance}$$

Using this model we can calculate the required distance as 46 m (the distance to speed up to 30 mph and to slow down from 30 mph).

A person solving this problem in real life may be able to set up some experimental bumps and observe the speeds of cars between them.

There are some ways in which this model could be revised or reformulated, for example:

- modelling the car as a rigid body that has length,
- looking for alternative models to describe how a car gains speed and slows down.

EXERCISE 1A

1 Gather some data on the lengths of cars and revise the solution to the speed bumps problem to take account of this factor.

2 Make a list of the assumptions that you might make to model the motion of a javelin.

3 If you were to model the motion of the pendulum in a grandfather clock:

(a) make a list of the assumptions that you might make,

(b) make a list of the variables that you might include in your model.

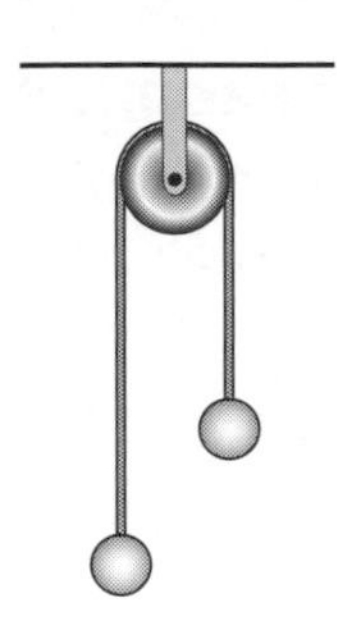

4 A string that passes over a pulley connects two objects of equal masses. The initial positions of the objects are shown in the diagram. Discuss the different predictions that you would obtain if you model:

(a) the pulley as smooth,

(b) the pulley as smooth and the string as light.

5 A student models a parachutist as a particle that does not experience air resistance. Suggest what predictions he might obtain.

Key point summary

1 The mathematical modelling cycle consists of: *p2*

- formulating the problem,
- obtaining a mathematical solution,
- validating and interpreting the solution,
- preparing a report.

2 Modelling requires assumptions to be made. *p2*

3 Particle and rigid body models are different as a particle has no size but has a mass, whereas a rigid body has size but does not change shape. *p3*

4 Terminology used in mechanics: *p3*

- smooth — no friction,
- rough — friction present,
- light — has no mass,
- inelastic — does not stretch,
- inextensible — does not stretch.

The mathematical modelling cycle

Formulation
- Specify problem
- Make assumptions
- Formulate a mathematical problem

Mathematical solution

Validation
- Interpret solution
- Be critical of solution
- Compare with reality

Report

CHAPTER 2

Kinematics in one dimension with constant acceleration

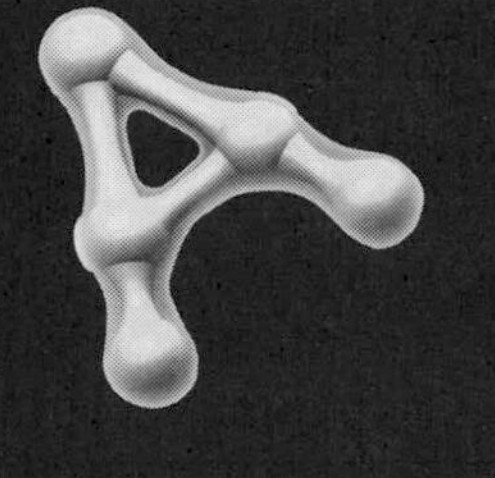

Learning objectives

After studying this chapter you should be able to:

- define displacement, velocity and acceleration
- understand and interpret displacement–time graphs and velocity–time graphs
- solve problems involving motion under constant acceleration using standard formulae
- solve problems involving motion under gravity.

2.1 Introducing kinematics

Kinematics is the study of motion and in this chapter we will study objects which move in one dimension only, this means that they move in a straight line. Examples are:

- cars, buses, or bikes, etc. on a straight road,
- objects dropped from the top of a cliff or tower,
- an athlete running in a 100 m race.

We will not, at this stage, consider what makes the objects move!

We shall also model most of the objects as particles so that we can ignore their size and shape and concentrate on their movement.

2.2 Displacement, velocity and acceleration

First, we have to define some words and symbols that we will use. In everyday language we use the terms **speed** and **distance** when talking about motion. **Distance** is how far we travel (miles, metres, etc.) and **speed** is how fast we go (miles per hour, metres per second, etc.). In the study of kinematics we need to be more precise and so we introduce **displacement** and **velocity**.

Displacement is based on the distance from a specific origin or reference point, but it also takes account of the direction in which the particle has moved.

The displacement may be 5 km north from a reference point or origin; or it may be given using positive or negative values relative to an origin, as shown in the diagram.

We decide that, relative to the point O, displacements to the right are positive and those to the left negative. Thus the displacement of P is $+3$ cm, and of Q -2 cm.

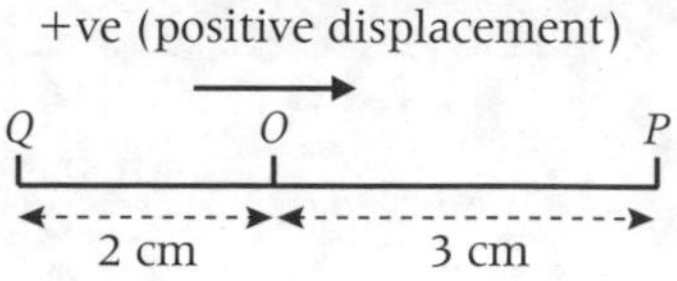

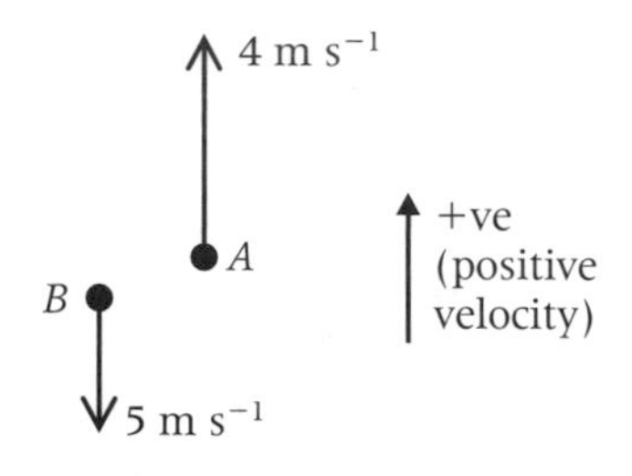

In mechanics the symbol used for displacement is s.

Velocity is defined similarly using the speed of an object together with the direction of the motion.

The velocity could be 6 mph going south-west from a reference point, or positive and negative values can be used.

In the diagram two particles are moving vertically. A is going up at 4 m s^{-1} and B is falling at 5 m s^{-1}. We choose upwards, say, as the positive direction so the velocity of A is $+4\text{ m s}^{-1}$ and B is -5 m s^{-1}.

The symbol used for velocity is v.

We often speak of **average speed** meaning the **constant speed** we could have travelled at in order to cover a journey in the same time. For example, if you travel a journey of 100 miles in 2 hours, your average speed is 50 mph; it is very unlikely that you could have driven at a constant 50 mph!

(Quantities that have size (magnitude) but no specific direction, such as speed and distance, are called **scalars**, but those with direction, like velocity and displacement, are called **vectors**. Vectors are studied in more detail in Chapter 3.)

Displacement–time graphs

As a body moves the displacement changes so that s is a function of time, t. A graph plotting s against t, is called a displacement–time graph. As an example the graph shows the displacement–time graph for the motion of a ball thrown in the air and falling back to the floor.

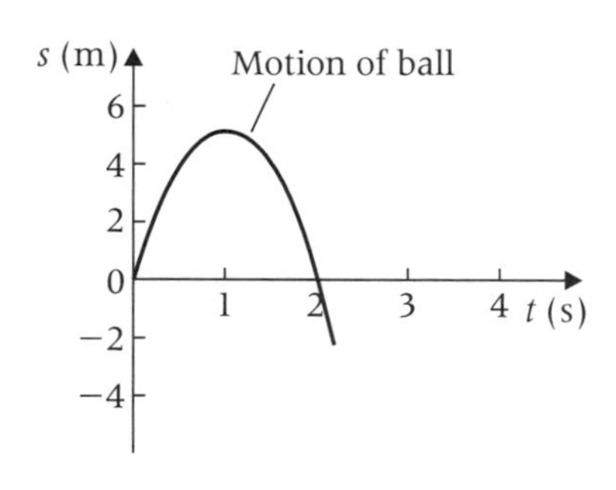

The motion of the ball can be described by looking at the graph. The ball starts at the origin and begins to move in the positive s direction (upwards) to a maximum height of 5 m above the point of release. It then falls to the floor which is 2 m below the point of release. It takes just over 2 s to hit the floor.

In another example of a displacement–time graph, the table shows the displacement (s) of a boy, who is running on a straight track, measured at 2-second intervals.

Time, t	0	2	4	6	8	10	12
Displacement, s	0	4	8	12	16	20	24

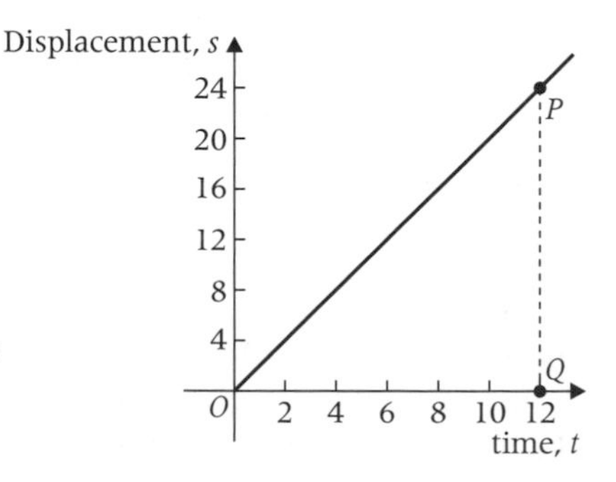

These values have been plotted on the graph. You notice that this graph is a straight line which tells us that the boy is running with a constant velocity.

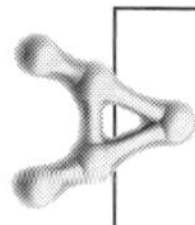

His velocity is calculated from the gradient of the graph

$$\text{gradient} = \frac{PQ}{OQ} = \frac{24}{12} = 2$$

So the boy runs at 2 m s^{-1}

Velocity–time graphs

If we know the velocity, v, at time, t, then we can draw a graph of velocity against time.

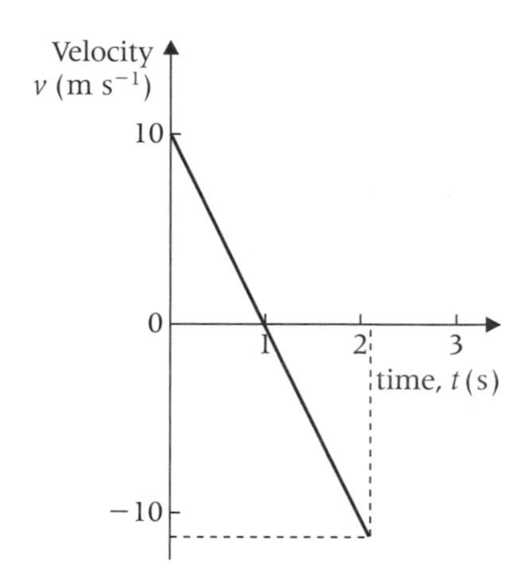

This velocity–time graph is for the motion of the ball whose displacement–time graph we saw above. The ball is thrown up into the air with a velocity of 10 m s^{-1} and this decreases to zero as the ball reaches the highest point, in about 1 second. The direction of motion is then reversed as the ball falls back to the floor, so for this part of the motion the velocity is negative. The ball hits the floor with a velocity of about -12 m s^{-1}

Here is another example of a velocity–time graph:

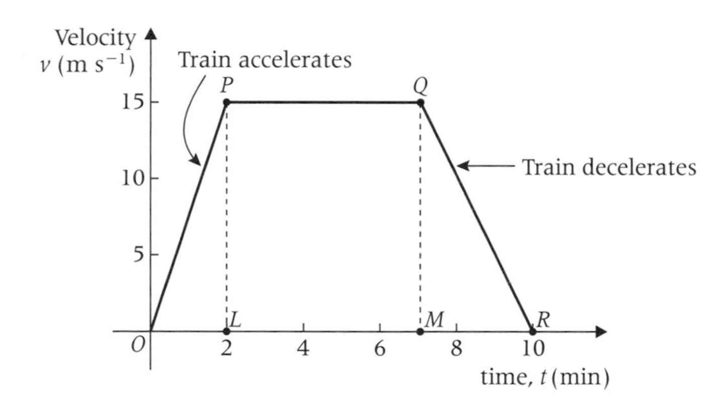

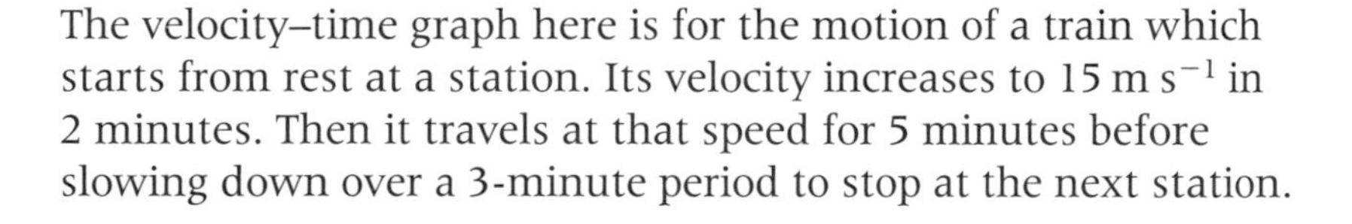

The velocity–time graph here is for the motion of a train which starts from rest at a station. Its velocity increases to 15 m s^{-1} in 2 minutes. Then it travels at that speed for 5 minutes before slowing down over a 3-minute period to stop at the next station.

Acceleration

Another familiar word used to describe the motion of cars, trains, bikes, and so on, in everyday language is **acceleration**. In the last two graphs above the velocity changes and the rate of change is called the acceleration.

Acceleration is defined as the rate at which the velocity is changing. Its units are 'metres per second per second' or m s^{-2}. So an acceleration of 5 m s^{-2} means that the velocity is increasing by 5 m s^{-1} every second.

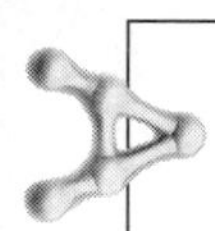

> We can calculate the acceleration from the gradient of a velocity–time graph. Look at the v–t graph of the train above.

During the first part of the motion:

$$\text{accleration} = \text{gradient of } OP$$
$$= \frac{15}{120}$$
$$= \frac{1}{8} \text{ or } 0.125 \text{ m s}^{-2}$$

Note that the time has been converted from minutes to seconds.

During the middle section the acceleration is zero.

In the third section of the train's journey it is slowing down; this is 'decelerating' or 'retarding' and the acceleration will have a negative value.

$$\text{acceleration} = \text{gradient of } QR$$
$$= \frac{-15}{180}$$
$$= -\frac{1}{12} \text{ or } -0.0833 \text{ m s}^{-2} \text{ (to 3 sf)}$$

Note that the symbol used for acceleration is a.

Displacement and the velocity–time graph

If a lorry moves at a constant speed of 15 m s^{-1} for 10 s, how far does it travel?
The distance that the lorry travels is simply given by

$$15 \times 10 = 150 \text{ m}$$

If we look at the velocity–time graph for the lorry, we can see that 150 is the area of the rectangle $OABC$. We refer to this as the 'the area under the graph'.

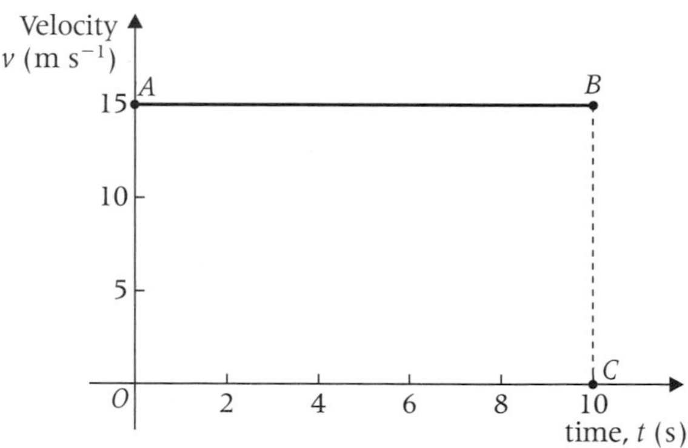

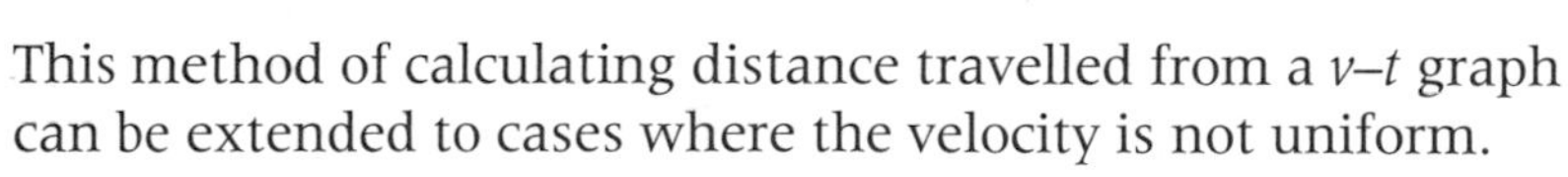

This method of calculating distance travelled from a v–t graph can be extended to cases where the velocity is not uniform.

Later the lorry accelerates from rest to 20 m s^{-1} in 8 s. The diagram shows its v–t graph.

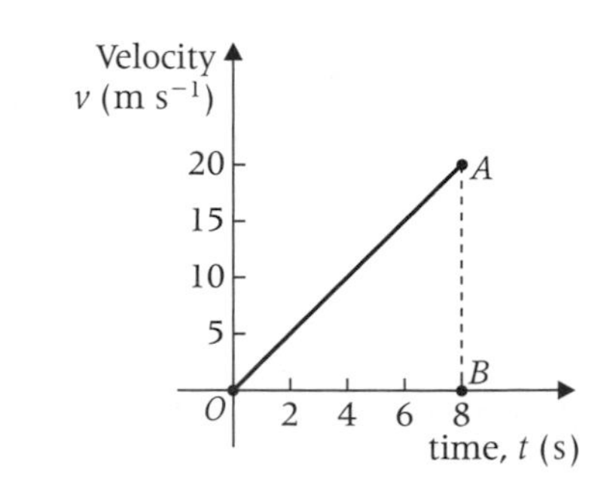

The area under this graph is the area of the triangle OAB, which can be calculated as

$$\text{Area} = \frac{20 \times 8}{2}$$
$$= 80$$

We can compare this with the average speed of the lorry over the 8-second period which would be 10 m s^{-1}. A lorry travelling at 10 m s^{-1} for 8 s would cover 80 m. Note that this distance is the same as our area.

(This example is not intended to be a 'proof', but merely an indication of how the method works.)

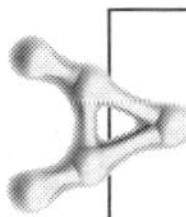

The area under a velocity–time graph represents the distance travelled.

Care must be taken with problems that include both positive and negative velocities. Consider the graph shown here. The shaded area marked A_1 represents a distance travelled in the positive direction, while the area marked A_2 represents a distance travelled in the negative direction. Comparing the sizes of the shaded areas indicates that a greater distance has been travelled in the positive direction than in the negative direction, so that the final displacement will be positive.

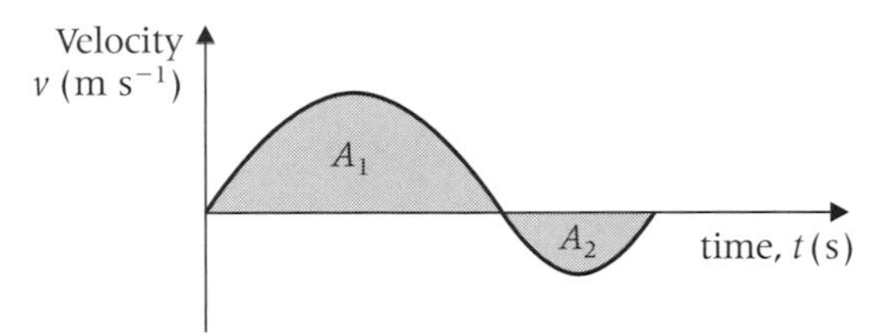

Worked example 2.1

Alongside a railway track there are marker posts, spaced at kilometre intervals. A train, travelling at constant velocity, is timed to take 2 minutes to travel from one post to the next. After passing the second post, the train slows down uniformly to stop in 0.4 km. Find

(a) the constant velocity of the train,

(b) the time it takes to stop,

(c) the acceleration of the train.

Solution

(There is a mixture of units in this question so we will use *seconds* for the time and *metres* for distance.)

First we make a sketch of the v–t graph, as shown. Note that BC is a straight line as the train slows down uniformly.

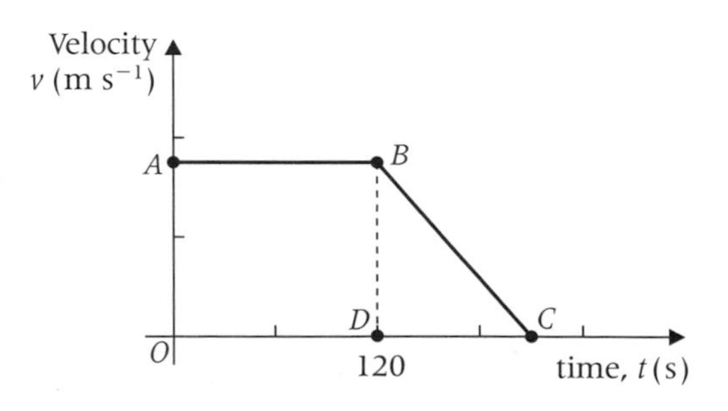

(a) The train travels 1000 m in 120 s.
The (constant) speed of the train

$$= \frac{1000}{120} = \frac{25}{3} = 8.33 \text{ m s}^{-1} \text{ (to 3 sf)}$$

(b) During the retardation, the train travels 400 m so the area of the triangle BDC must be 400 units, hence

$$\frac{BD \times DC}{2} = 400$$

$$DC = 96$$

Thus the train slows down for 96 s.

(c) The acceleration of the train is the gradient of BC.

$$a = \frac{-8.33}{96}$$
$$= -0.0868 \text{ m s}^{-2} \text{ (to 3 sf)}$$

Worked example 2.2

A cyclist rides along a straight road from X to Y. He starts from rest at X and accelerates uniformly to reach a speed of 10 m s^{-1} in 8 s. He travels at this speed for 20 s and then decelerates uniformly to stop at Y. If the whole journey takes 40 s, sketch a velocity–time graph for the journey.

Use the graph to find:

(a) the initial acceleration,

(b) the acceleration on the final stage,

(c) the total distance travelled.

Solution

The v–t graph is shown below.

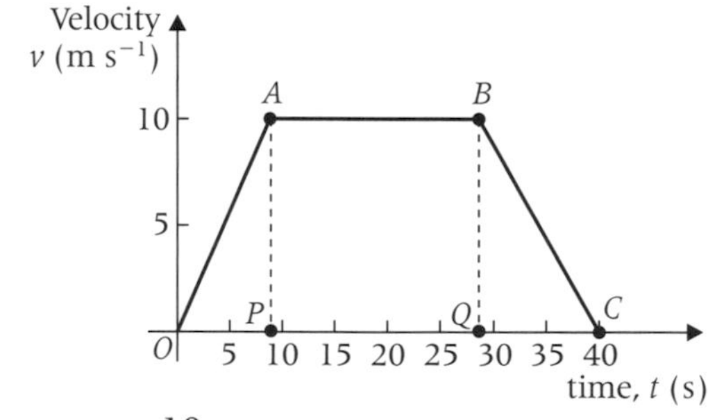

(a) Acceleration $= \frac{10}{8} = 1.25 \text{ m s}^{-2}$

(b) Acceleration $= \frac{-10}{12}$

$$= -\frac{5}{6} \text{ m s}^{-2}$$

We could say that the acceleration is $-\frac{5}{6} \text{ m s}^{-2}$ or that the cyclist decelerates at $\frac{5}{6} \text{ m s}^{-2}$.

(c) Total distance = Area APO + Area $ABQP$ + Area BQC

$$= \frac{8 \times 10}{2} + 20 \times 10 + \frac{12 \times 10}{2}$$

$$= 40 + 200 + 60 = 300 \text{ m.}$$

EXERCISE 2A

1 The graph shows how the velocity of a car changes during a short journey. Find the distance travelled by the car and the acceleration on each stage of its journey.

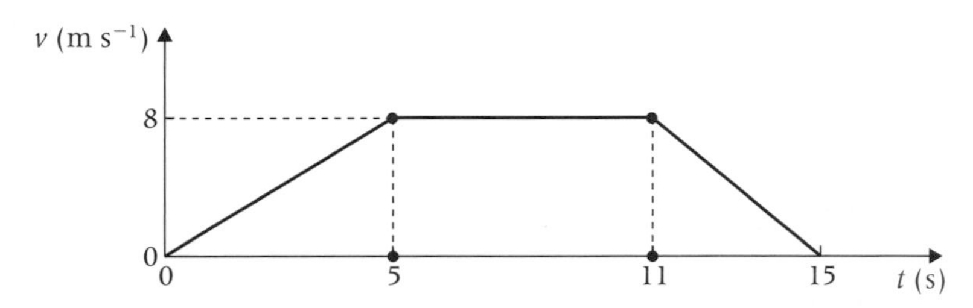

2 This graph shows how the velocity of a cyclist changes over a short period of time. Find the total distance travelled by the cyclist.

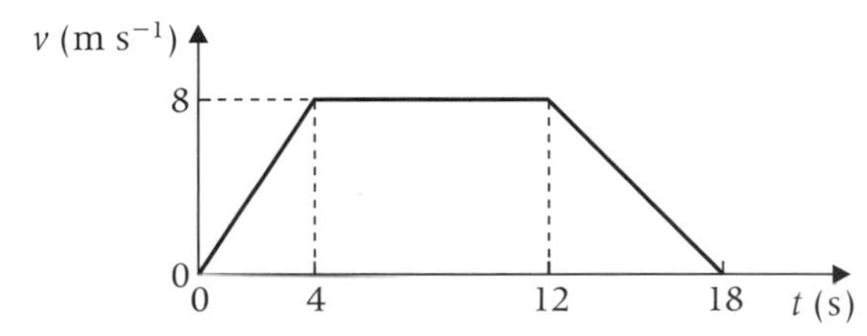

3 Discuss the motion represented by each of the displacement–time graphs shown here.

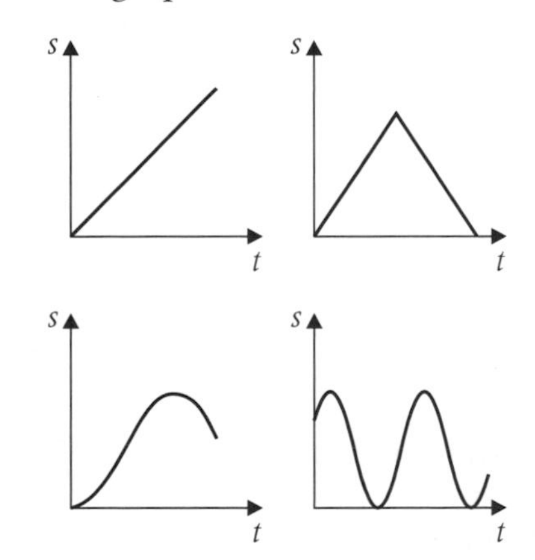

4 Sketch displacement–time and velocity–time graphs for the following.

(a) A car that starts from rest and increases its velocity steadily to 10 m s^{-1} in 5 s. The car holds this velocity for another 10 s and then slows steadily to rest in a further 10 s.

(b) A ball that is dropped on to a horizontal floor from a height of 3 m. The ball bounces several times before coming to rest.

(c) A person who jumps out of a balloon and falls until the parachute opens. The person then glides steadily to the ground.

5

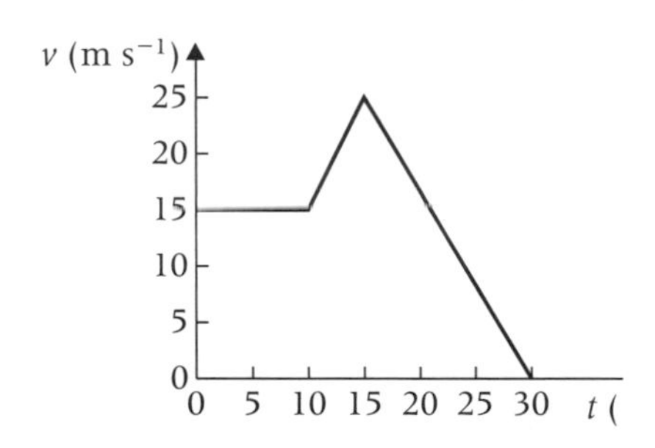

This velocity–time graph illustrates the motion of an object. Calculate the acceleration for each of the following intervals:

(a) $0 < t < 10$,

(b) $10 < t < 15$,

(c) $15 < t < 30$,

(d) Calculate the displacement of the object over the 30 s.

6 A car accelerates at 2 m s^{-2} from rest until it reaches a speed of 16 m s^{-1}. It then travels at this speed for 30 s, before slowing down and stopping in a further 5 s.

Find the total distance travelled by the car. [A]

7

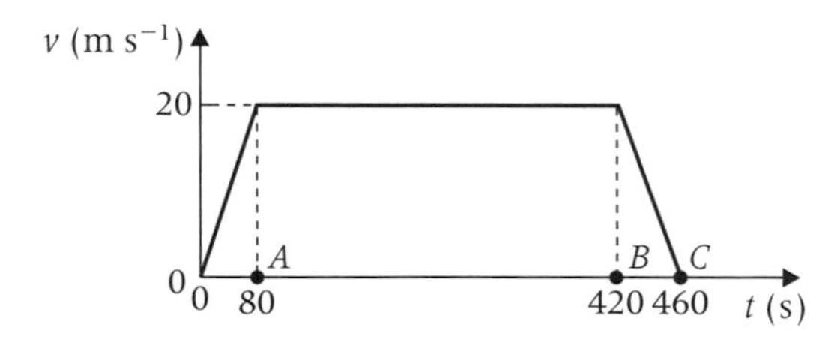

The diagram shows the velocity–time graph for a train which travels from rest in one station to rest at the next station. For each of the time intervals *OA*, *AB* and *BC*, state the value of the train's acceleration.

Calculate the distance between the stations. [A]

8 A train is travelling at a constant speed of 40 m s^{-1}, when the driver sees a warning light. Over the next 1000 m the speed of the train drops to 20 m s^{-1}. The train travels at this speed for

5 minutes. The speed returns to 40 m s^{-1} after a further 5 minutes. Assume that the acceleration of the train is constant on each stage of its journey.

(a) Find the total distance travelled by the train, while its speed is less than its normal operating speed of 40 m s^{-1}.

(b) The train **would normally** have travelled this distance at a constant 40 m s^{-1}. Find the time by which it was delayed. [A]

9 A tram travelling along a straight track starts from rest and accelerates uniformly for 15 s. During this time it travels 135 m. The tram now maintains a constant speed for a further 1 minute. It is finally brought to rest decelerating uniformly over a distance of 90 m. Calculate the tram's acceleration and deceleration during the first and last stages of the journey. Also find the time taken and the distance travelled for the whole journey. [A]

10 A train travelling at 50 m s^{-1} applies its brakes on passing a yellow signal at a point A and decelerates uniformly, with a deceleration of 1 m s^{-2}, until it reaches a speed of 10 m s^{-1}. The train then travels for 2 km at the uniform speed of 10 m s^{-1} before passing a green signal. On passing the green signal the train accelerates uniformly, with acceleration 0.2 m s^{-2}, until it finally reaches a speed of 50 m s^{-1} at a point B. Find the distance AB and the time taken to travel that distance. [A]

11 Two sprinters compete in a 100 m race, crossing the finishing line together after 12 s. The two models, A and B, as described below, are models for the motions of the two sprinters.

Model A. The sprinter accelerates from rest at a constant rate for 4 s and then travels at a constant speed for the rest of the race.

Model B. The sprinter accelerates from rest at a constant rate until reaching a speed of 9 m s^{-1} and then travels at this speed for the rest of the race.

(a) For model A, find the maximum speed and the initial acceleration of the sprinter.

(b) For model B, find the time taken to reach the maximum speed and the initial acceleration of the sprinter.

(c) Sketch a distance–time graph for each of the two sprinters on the same set of axes. Describe how the distance between the two sprinters varies through the race. [A]

12 A car travels a total distance of 430 m in a time of 25 s. During this time, the car accelerates from rest at 5 m s^{-2} for 4 s, then travels at a constant speed and finally slows down, with constant deceleration, until it stops.

(a) Find the distance travelled by the car in the first 4 s and the speed of the car at the end of this time.

(b) Find the time for which the car travels at a constant speed and the deceleration during the final stage of the car's motion. [A]

2.3 Motion under constant acceleration

There are several simple formulae which can be used to solve problems that involve motion under **constant** (or **uniform**) acceleration.

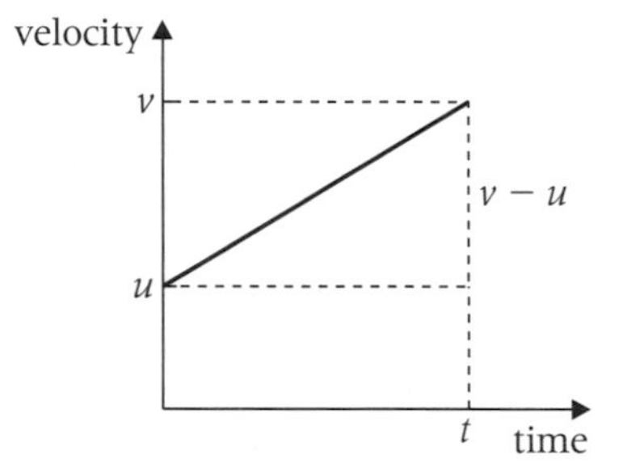

The diagram is a velocity–time graph for the motion of an object with initial velocity u and final velocity v after t seconds has elapsed.

The gradient of the line is equal to the acceleration and is calculated from the expression

$$\frac{v-u}{t}$$

Hence

$$a = \frac{v-u}{t}$$

which can be rewritten as

$$v = u + at.$$

The area under the velocity–time graph is equal to the displacement of the object. Using the rule for the area of a trapezium gives

$$s = \frac{1}{2}(u+v)t$$

Worked example 2.3

A motorbike accelerates at a constant rate of 3 m s^{-2}. Calculate:

(a) the time taken to accelerate from 18 km h^{-1} to 45 km h^{-1},

(b) the distance, in metres, covered in this time.

Solution

We can use the equation $v = u + at$ to find the time and then the equation $s = \frac{1}{2}(u + v)t$ to find the distance travelled. But first the units for speed must be converted to m s^{-1}.

$$18 \text{ km h}^{-1} = \frac{18 \times 1000}{3600} = 5 \text{ m s}^{-1}$$

and similarly

$$45 \text{ km h}^{-1} = 12.5 \text{ m s}^{-1}$$

(a) Using $v = u + at$, with $u = 5$, $v = 12.5$ and $a = 3$ gives

$$12.5 = 5 + 3t$$

$$t = \frac{7.5}{3} = 2.5 \text{ s}$$

(b) Using $s = \frac{1}{2}(u + v)t$ with $u = 5$, $v = 12.5$ and $t = 2.5$ gives

$$= \frac{1}{2}(12.5 + 5) \times 2.5$$

$$= 21.875 \text{ m}$$

Two more useful formulae

We can write the equation $v = u + at$ in the form

$$t = \frac{v - u}{a}$$

and substitute it into $s = \frac{1}{2}(u + v)t$, then

$$s = \frac{1}{2}(u + v)\left(\frac{v - u}{a}\right)$$

$$= \frac{v^2 - u^2}{2a}$$

which we can rearrange to give

$$v^2 = u^2 + 2as$$

Similarly, using $v = u + at$ to substitute for v in the equation

$$s = \frac{1}{2}(u + v)t$$

gives

$$s = \frac{(u + u + at)t}{2}$$

or

$$s = ut + \frac{1}{2}at^2$$

This gives four different constant acceleration formulae. The appropriate one can be selected to fit the data available in the problem that you have to solve.

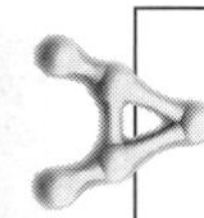

When using these formulae it is important to remember that they only apply to cases where the acceleration is constant or can be assumed to be constant.

Worked example 2.4

A car accelerates from a velocity of $16\,\mathrm{m\,s^{-1}}$ to a velocity of $40\,\mathrm{m\,s^{-1}}$ in a distance of 500 m. Find the acceleration of the car.

Solution

Using the equation $v^2 = u^2 + 2as$, with $u = 16$, $v = 40$ and $s = 500$ gives

$$40^2 = 16^2 + 2 \times a \times 500$$

$$a = \frac{1600 - 256}{1000} = 1.344\,\mathrm{m\,s^{-2}}$$

Worked example 2.5

A car decelerates from a velocity of $36\,\mathrm{m\,s^{-1}}$. The magnitude of the deceleration is $3\,\mathrm{m\,s^{-2}}$. Calculate the time required to travel a distance of 162 m.

Solution

When an object is slowing down (decelerating) we can use the constant acceleration equations, but with a negative value for a. In this case $a = -3$.

We require the time, t s to travel 162 m.

Using the constant acceleration equation $s = ut + \frac{1}{2}at^2$, with $s = 162$, $u = 36$ and $a = -3$ gives

$$162 = 36t + \tfrac{1}{2}(-3)t^2$$

Rearranging this gives this quadratic equation:

$$1.5t^2 - 36t + 162 = 0$$

Dividing by 1.5 gives

$$t^2 - 24t + 108 = 0$$

And factorising gives

$$(t - 6)(t - 18) = 0$$

so that

$$t = 6 \text{ or } t = 18$$

The first answer, of 6 s, is the required one.

The second answer would give the time that the displacement was 162 m for the second time. In this case the car would be moving in the opposite direction.

Worked example 2.6

A cyclist is initially travelling at $10\ \text{m s}^{-1}$, when she applies her brakes. Assume that her acceleration remains constant at $-0.8\ \text{m s}^{-2}$ until she stops. Find the distance that she travels before stopping and the time that it takes her to stop.

Solution

To find the distance that she travels use the formula $v^2 = u^2 + 2as$, with $v = 0$, $u = 10$ and $a = -0.8$, which gives

$$0^2 = 10^2 + 2 \times (-0.8)s$$

$$0 = 100 - 1.6s$$

$$s = \frac{100}{1.6} = 62.5\ \text{m}$$

Now the time taken to stop can be found using the formula $s = \frac{1}{2}(u + v)t$, with $s = 62.5$, $u = 10$ and $v = 0$, to give

$$s = \frac{1}{2}(u + v)t$$

$$62.5 = \frac{1}{2}(10 + 0)t$$

$$62.5 = 5t$$

$$t = 12.5\ \text{s}$$

EXERCISE 2B

1 A car accelerates at $2\ \text{m s}^{-2}$ from rest for 10 s.

(a) Find the distance travelled by the car and the speed it reaches.

(b) After the 10 s its acceleration changes to $0.5\ \text{m s}^{-2}$ and then remains constant for a further 5 s. Find the speed of the car and the total distance that it has travelled at the end of the 15 s.

2 A car accelerates from $10\ \text{m s}^{-1}$ to $20\ \text{m s}^{-1}$ as it travels 500 m.

(a) Find the acceleration of the car.

(b) Find the time taken by the car to travel the 500 m.

3 A car accelerates uniformly from 5 m s^{-1} to 12 m s^{-1}, in a 10-second period of time.

(a) Find the acceleration of the car.

(b) Find the distance travelled by the car.

4 A lift rises from rest, accelerating at a constant rate until it reaches a speed of 1.6 m s^{-1} after 8 s.

(a) Find the acceleration of the lift.

(b) The lift continues to accelerate for a further 2 s. Find the distance that the lift has now risen.

(c) The lift then shows down, at a constant rate, and stops after a further 5 s. Find the total distance travelled by the lift.

5 A car accelerates uniformly from a speed of 50 kph to a speed of 80 kph in 20 s.

Calculate the acceleration in m s^{-2}.

6 For the car in Question **5**, calculate the distance travelled during the 20 s.

7 A van travelling at 40 mph skids to a halt in a distance of 15 m. Find the acceleration of the van and the time taken to stop, assuming that the deceleration is uniform. (Assume 1 mile = 1600 m.)

8 A train signal is placed so that a train can decelerate uniformly from a speed of 96 km h^{-1} to come to rest at the end of a platform. For passenger comfort the deceleration must be no greater than 0.4 m s^{-2}. Calculate

(a) the shortest distance the signal can be from the platform,

(b) the shortest time for the train to decelerate.

9 A rocket is travelling with a velocity of 80 m s^{-1}. The engines are switched on for 6 s and the rocket accelerates uniformly at 40 m s^{-2}. Calculate the distance travelled over the 6 s.

10 The world record for the men's 60 m race was 6.41 s.

(a) Assuming that the race was carried out under constant acceleration, calculate the acceleration of the runner and his speed at the end of the race.

(b) Now assume that in a 100 m race the runner accelerates for the first 60 m and completes the race by running the next 40 m at the speed you calculated in **(a)**.

Calculate the time for the athlete to complete the race.

11 The world record for the men's 100 m was 9.83 s. Assume that the last 40 m was run at constant speed and that the acceleration during the first 60 m was constant.

(a) Calculate this speed.

(b) Calculate the acceleration of the athlete.

12 Telegraph poles, 40 m apart, stand alongside a railway line. The times taken for a locomotive to pass the two gaps between three consecutive poles are 2.5 s and 2.3 s, respectively. Calculate the acceleration of the train and the speed past the first post.

13 A set of traffic lights covers road repairs on one side of a road in a 30 mph speed limit area. The traffic lights are 80 m apart so time must be allowed to delay the light changing from green to red. Assuming that a car accelerates at 2 m s^{-2} what is the least this time delay should be?

14 A train starts from rest and moves with constant acceleration $\frac{1}{3} \text{ m s}^{-2}$ for 2 minutes. For the next 4 minutes the train moves with zero acceleration, after which a uniform retardation of 2 m s^{-2} brings it to rest. Find the total distance travelled by the train from starting to stopping. [A]

15 As a train leaves a station it accelerates, from rest, at 0.8 m s^{-2} for 30 s, travels at a constant speed for the next 5 minutes and then slows down, stopping in 20 s at a second station.

(a) Find the maximum speed of the train.

(b) Find the distance travelled by the train between the stations, clearly stating any assumptions that you have made. [A]

16 As a lift moves upwards from rest it accelerates at 0.8 m s^{-2} for 2 s, then travels 4 m at constant speed and finally slows down, with a constant deceleration, stopping in 3 s.

Find the total distance travelled by the lift and the total time taken. [A]

17 Two cars A and B are initially at rest side by side. A starts off on a straight track with an acceleration of 2 m s^{-2}. Five seconds later B starts off on a parallel track to A, with acceleration 3.125 m s^{-2}.

(a) Calculate the distance travelled by A after 5 seconds.

(b) Calculate the time taken for B to catch up A.

(c) Find the speeds of A and B at that time.

18 Two cars are initially 36 m apart travelling in the same direction along a straight, horizontal road. The car in front is initially travelling at $10\ \text{m s}^{-1}$, but decelerating at $2\ \text{m s}^{-2}$. The other car travels at a constant $15\ \text{m s}^{-1}$.

(a) Model the cars as particles. By finding the distance travelled by each car after t s, show that the distance between the two cars is $36 - 5t - t^2$ m. Find when they would collide if neither car takes avoiding action.

(b) Would it be necessary to revise your answers to part **(a)** if the cars were not modelled as particles? Give reasons to support your answer. [A]

19 Two humps are to be installed on a road to prevent traffic reaching speeds of greater than $12\ \text{m s}^{-1}$ between the humps. Assume that:

I the speed of cars when they cross the humps is effectively zero,

II after crossing a hump they accelerate at $3\ \text{m s}^{-2}$ until they reach a speed of $12\ \text{m s}^{-1}$,

III as soon as they reach a speed of $12\ \text{m s}^{-1}$ they decelerate at $6\ \text{m s}^{-2}$ until they stop.

(a) A simple model ignores the lengths of the cars. Use this to find the distance between the humps.

(b) One factor that has not been taken into account is the length of the cars. Revise your answer to **(a)** to take this into account, giving your answer to the nearest metre. You must state clearly any assumptions that you make. [A]

2.4 Motion under gravity

For many centuries it was believed that:

(a) heavier bodies fell faster than light ones, and

(b) the speed of a falling body was constant all through its motion.

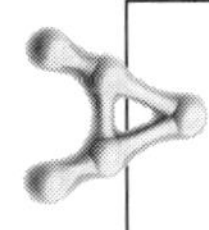

Galileo Galilei (1564–1642) was the first person to state clearly (and to demonstrate) that all objects fall with the same acceleration.

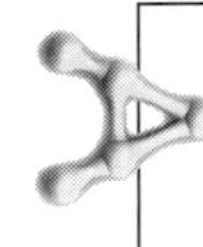

Modern scientific instruments determine the acceleration of falling bodies as values in the region of $9.81\ \text{m s}^{-2}$, although the value varies slightly at different places on the earth's surface, and at different altitudes. The symbol used to represent this 'acceleration due to gravity' is g.

The value of g is sometimes approximated to $10\,\text{m s}^{-2}$, but in this book we shall normally use $9.8\,\text{m s}^{-2}$ unless stated otherwise. Since this acceleration acts towards the earth's surface, its sign must always be opposite to that of any velocities that are upwards.

When solving problems involving motion under gravity (ignoring any air resistance at this stage) the formulae for motion with constant acceleration may be used.

Worked example 2.7

A ball is projected vertically upwards with an initial speed of $30\,\text{m s}^{-1}$. Calculate the maximum height reached.

Solution

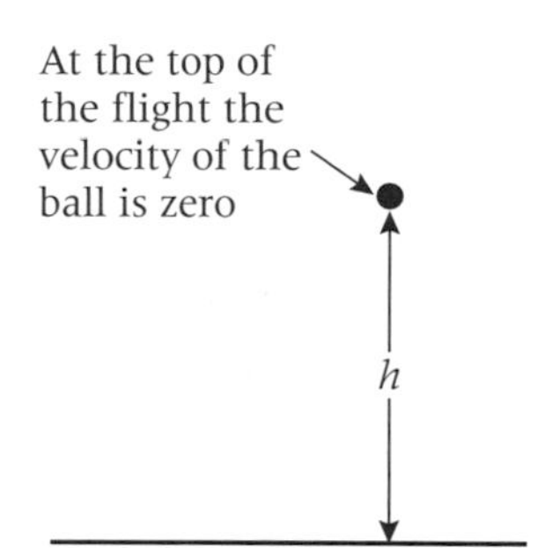

At the top of the ball's flight, its speed will be zero.

Take the upwards direction as positive, so the acceleration will be

$$a = -g = -9.8\,\text{m s}^{-2}.$$

Using $v^2 = u^2 + 2as$ gives

$$0 = 30^2 - 2 \times 9.8 \times h$$

where h is the maximum height reached.

Thus

$$h = \frac{900}{19.6} = 45.9\,\text{m (to 3 sf)}.$$

Worked example 2.8

A stone is fired vertically upwards with an initial speed of $10\,\text{m s}^{-1}$ from a catapult. Calculate the interval between the two times when the ball is 5 m above the point of release.

Solution

The stone is at a height of 5 m on its upward journey and again on the way down. The neatest way to solve the problem is as follows.

Using $s = ut + \frac{1}{2}at^2$ with

$$a = -g = -9.8\,\text{m s}^{-2}$$

$$5 = 10t - 0.5 \times 9.8 \times t^2$$

where t s is the time passed since the stone was thrown up.

$$4.9t^2 - 10t + 5 = 0$$

Using the quadratic equation formula

$$t = \frac{10 \pm \sqrt{2}}{9.8} = 0.876 \text{ or } 1.165$$

the required time interval is $1.165 - 0.876 = 0.289$ s.

EXERCISE 2C

1 A ball is dropped from rest at a height of 2 m. Find the time that the ball takes to fall to the ground, if it is:

(a) on earth,

(b) on the moon, where $g = 1.6 \text{ m s}^{-2}$.

2 A ball, that is initially at rest, falls from a height of 3 m to the ground.

(a) Find the time that the ball takes to fall this distance.

(b) Find the speed of the ball when it hits the ground.

3 A ball is thrown upwards with an initial speed of 14.7 m s^{-1} from a height of 1 m.

(a) Find the time that it takes the ball to reach its maximum height.

(b) Find the maximum height of the ball.

(c) Find the speed of the ball when it hits the ground.

4 A rocket rises from ground level to a height of 100 m in 10 s. Assume that the acceleration of the rocket is constant and that it starts at rest.

(a) Find the acceleration of the rocket and its speed at a height of 100 m.

After these 10 s the rocket runs out of fuel, but continues to move vertically under the influence of gravity.

(b) Find the maximum height of the rocket.

5 The diagram shows three positions of a ball which has been thrown upwards with a velocity of $u \text{ m s}^{-1}$.

Position A is the initial position.

Position B is halfway up.

Position C is at the top of the motion.

(a) Copy the diagram and for each position put on arrows where appropriate to show the direction of the velocity.

(b) On the same diagram put on arrows to show the direction of the acceleration.

6 A ball is dropped on to level ground from a height of 20 m.

(a) Calculate the time taken to reach the ground.

The ball rebounds with half the speed it strikes the ground.

(b) Calculate the time taken to reach the ground a second time.

7 A stone is thrown down from a high building with an initial velocity of 4 m s^{-1}. Calculate the time required for the stone to drop 30 m and its velocity at this time.

8 A ball is thrown vertically upwards from the top of a cliff which is 50 m high. The initial velocity of the ball is 25 m s^{-1}. Calculate the time taken to reach the bottom of the cliff and the velocity of the ball at that instant.

9 One stone is thrown upwards with a speed of 2 m s^{-1} and another is thrown downwards with a speed of 2 m s^{-1}. Both are thrown at the same time from a window 5 m above ground level.

(a) Which hits the ground first?

(b) Which is travelling fastest when it hits the ground?

(c) What is the total distance travelled by each stone?

10 A ball is thrown vertically upwards with an initial velocity of 30 m s^{-1}. One second later, another ball is thrown upwards with an initial velocity of $u\text{ m s}^{-1}$. The particles collide after a further 2 s. Find the value of u.

11 When a ball hits the ground it rebounds with half of the speed that it had when it hit the ground. If the ball is dropped from a height h, calculate the height to which it rebounds.

12 A small canister is attached to a helium-filled balloon and released from rest at ground level. After 4 s it is moving vertically upwards at 6 m s^{-1}.

(a) Find the height of the balloon and canister after 4 s, stating clearly any assumptions that you make.

When the balloon reaches a height of 27 m it bursts.

(b) Find the maximum height reached by the canister. [A]

13 A tennis ball is hit so that it moves vertically downwards from a height of 1 m with an initial speed of 5 m s^{-1}. When it hits the ground it rebounds vertically with half the speed it had when it hit the ground.

(a) Find the height to which it rebounds.

(b) State whether this is likely to be an underestimate or an overestimate, giving reasons to support your answer. [A]

Key point summary

Formulae to learn

$v = u + at$

$s = \frac{1}{2}(u + v)t$

$v^2 = u^2 + 2as$

$s = ut + \frac{1}{2}at^2$

1 The gradient of a displacement–time graph gives the velocity. *pp8–9*

2 The gradient of a velocity–time graph gives the acceleration. *p10*

3 The area under a velocity–time graph can be used to find the displacement. *p11*

4 Only use the constant acceleration formulae when the acceleration is constant or can be assumed to be constant. *p18*

5 All objects accelerate in the same way when falling under the influence of gravity alone. *p22*

6 The acceleration due to gravity is 9.8 m s^{-2}. *p22*

Test yourself

What to review

1 A car accelerates uniformly from rest to 20 m s^{-1} in 25 s. It travels at this speed for 1.5 minutes and then slows down, stopping after a further 35 s. — *Section 2.2*

(a) Draw a velocity–time graph and use it to find the total distance travelled by the car.

(b) Calculate the acceleration of the car on each stage of its journey.

2 The velocity of a car increases from 5 m s^{-1} to 25 m s^{-1} as it travels a distance of 100 m. Assume that the acceleration of the car is constant. — *Section 2.3*

(a) Find the acceleration of the car.

(b) Find the speed of the car when it has travelled 50 m.

(c) Find the time it takes for the car to travel the 100 m.

3 A stone is thrown upwards from a height of 2 m above ground level. It reaches a maximum height of 5 m above ground level. — *Section 2.4*

(a) Find the initial velocity of the stone.

(b) Find the velocity of the stone when it hits the ground.

(c) How long is the stone in the air?

Test yourself ANSWERS

1 **(a)** 2400 m, **(b)** $0.8\ \text{m s}^{-2}$, $0\ \text{m s}^{-2}$, $-\frac{4}{7}\ \text{m s}^{-2}$.

2 **(a)** $3\ \text{m s}^{-2}$, **(b)** $18.0\ \text{m s}^{-1}$, **(c)** $6\frac{2}{3}$ s.

3 **(a)** $7.67\ \text{m s}^{-1}$, **(b)** $9.90\ \text{m s}^{-1}$, **(c)** 1.79 s.

CHAPTER 3

Kinematics in two and three dimensions with constant acceleration

Learning objectives

After studying this chapter you should be able to:

- plot and interpret paths given a position vector
- write positions, velocities and accelerations in the form $x\mathbf{i} + y\mathbf{j} + z\mathbf{k}$ or $x\mathbf{i} + y\mathbf{j}$
- find magnitudes and directions of vectors
- apply and use the constant acceleration equations in two or three dimensions.

3.1 Introduction

In this chapter the idea of using constant acceleration is extended into two and three dimensions. Before doing this, we need to be able to describe motion in more than one dimension. This is done using vectors, which we will consider first.

3.2 Describing motion in two and three dimensions

We will first consider two dimensions and then extend the ideas to three dimensions.

The first key step is to define an origin or reference point. The position of an object is then described relative to this point. Often the origin will be the initial position of an object.

Secondly, we define two perpendicular unit vectors. These have length and are directed at right angles to each other.

The diagram at the top of the next page shows an origin O and two unit vectors $\mathbf{i}$ and $\mathbf{j}$.

In addition the diagram shows some points. We can describe the position of each point by using the unit vectors $\mathbf{i}$ and $\mathbf{j}$. For example the position of the point A can be written as:

$$\mathbf{r}_A = 4\mathbf{i} + 3\mathbf{j}$$

Similarly

$$\mathbf{r}_B = 5\mathbf{i} - 2\mathbf{j}$$
$$\mathbf{r}_C = -5\mathbf{i} - 3\mathbf{j}$$
$$\mathbf{r}_D = -3\mathbf{i} + 2\mathbf{j}$$

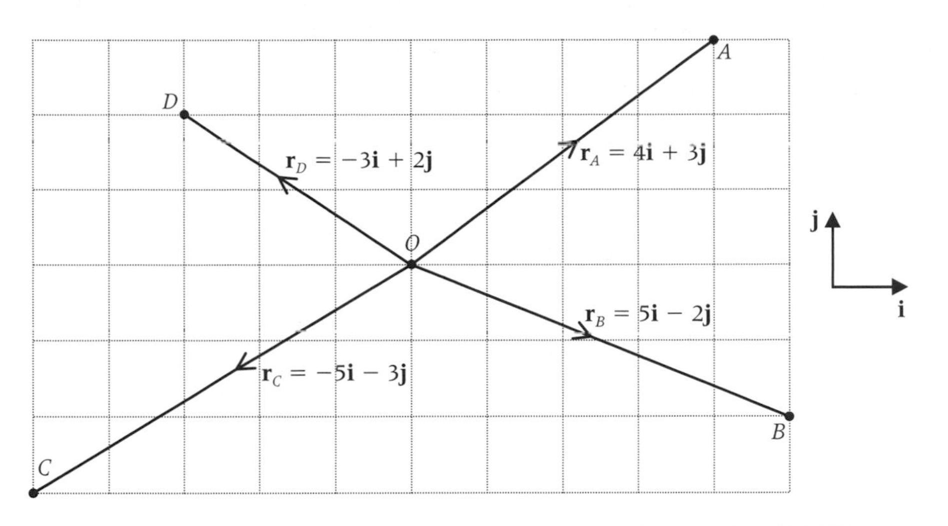

3

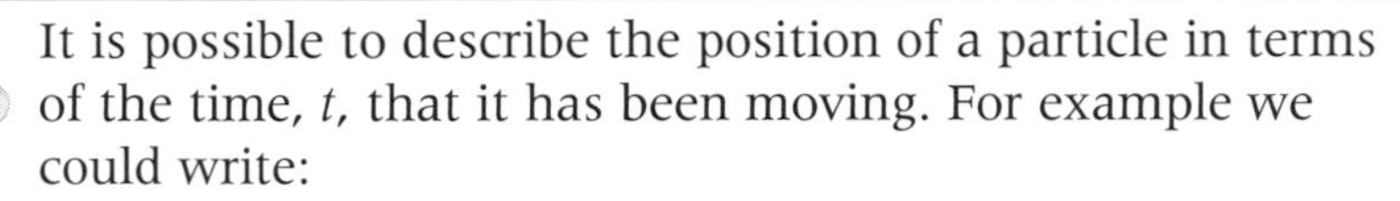

It is possible to describe the position of a particle in terms of the time, t, that it has been moving. For example we could write:

$$\mathbf{r} = (4t + 3)\mathbf{i} + (t^2 - 4t)\mathbf{j}$$

Then the position vector $\mathbf{r}$ can be found for any value of t and the path of the object can be drawn or described.

For two-dimensional motion we usually write:

$$\mathbf{r} = x(t)\mathbf{i} + y(t)\mathbf{j}$$

where $\mathbf{i}$ and $\mathbf{j}$ lie in the plane that the motion takes place in. If this is a vertical plane it is the normal convention that $\mathbf{i}$ is horizontal and $\mathbf{j}$ is vertical. The diagram shows the position of a particle.

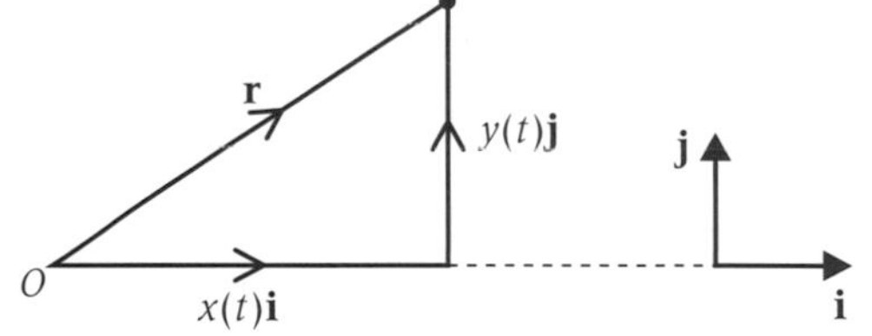

$$\mathbf{r} = (4t + 3)\mathbf{i} + (t^2 - 4t)\mathbf{j}$$

Then the position vector $\mathbf{r}$ can be found for any value of t and the path of the object can be drawn or described.

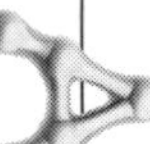

For three-dimensional motion we usually write:

$$\mathbf{r} = x(t)\mathbf{i} + y(t)\mathbf{j} + z(t)\mathbf{k}$$

where $\mathbf{i}$ and $\mathbf{j}$ lie in a horizontal plane and $\mathbf{k}$ is a vertical unit vector. Note that these three unit vectors are all perpendicular as shown in the diagram.

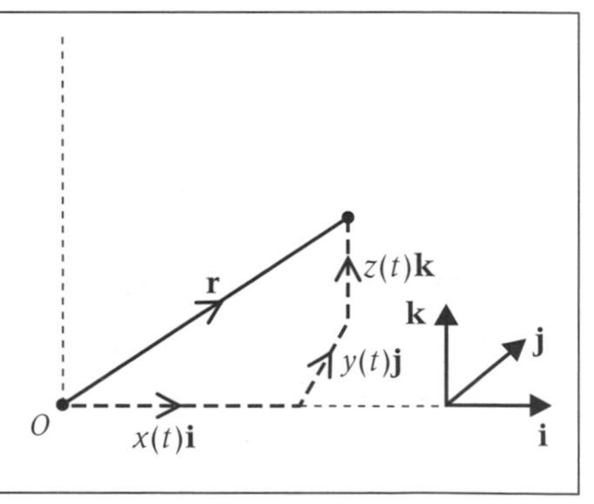

Note that $\mathbf{r}$ is referred to as the position vector of the object that is moving.

Worked example 3.1

A ball is thrown so that its position, in metres, at time t seconds is given by

$$\mathbf{r} = 6t\mathbf{i} + (1 + 8t - 5t^2)\mathbf{j}$$

where **i** and **j** are horizontal and vertical unit vectors, respectively.

(a) Find the position of the particle when $t = 0, 0.5, 1, 1.5$ and 2 s.

(b) Plot the positions in part **(a)** and draw the path of the particle.

Solution

(a) The table below shows how the values of t are substituted.

t	$\mathbf{r}$
0	$6 \times 0\mathbf{i} + (1 + 8 \times 0 - 5 \times 0^2)\mathbf{j} = 0\mathbf{i} + 1\mathbf{j}$
0.5	$6 \times 0.5\mathbf{i} + (1 + 8 \times 0.5 - 5 \times 0.5^2)\mathbf{j} = 3\mathbf{i} + 3.75\mathbf{j}$
1	$6 \times 1\mathbf{i} + (1 + 8 \times 1 - 5 \times 1^2)\mathbf{j} = 6\mathbf{i} + 4\mathbf{j}$
1.5	$6 \times 1.5\mathbf{i} + (1 + 8 \times 1.5 - 5 \times 1.5^2)\mathbf{j} = 9\mathbf{i} + 1.75\mathbf{j}$
2	$6 \times 2\mathbf{i} + (1 + 8 \times 2 - 5 \times 2^2)\mathbf{j} = 12\mathbf{i} - 3\mathbf{j}$

(b) These positions can then be plotted and a curve drawn to show the path of the ball.

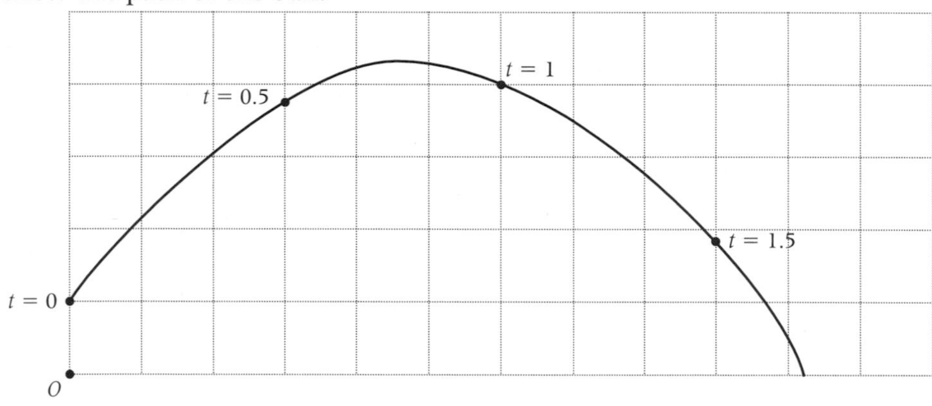

If the origin is at ground level we can note, from the diagram, that:

(i) the ball is thrown from a height of 1 m

(ii) the ball reaches a maximum height of 4.2 m

(iii) the horizontal distance travelled by the ball until it hits the ground is 10.3 m.

The ball has hit the ground before $t = 2$ s so this isn't shown on the plot.

Worked example 3.2

Two boats move on a pond. They both start at the same place. One follows a curved path and the other travels along a straight line. The positions, in metres, of the boats, at time t s are given by:

$$\mathbf{r}_A = 2t\mathbf{i} + 4t\mathbf{j}$$

$$\mathbf{r}_B = (8t - t^2)\mathbf{i} + 4t\mathbf{j}$$

where **i** and **j** are unit vectors directed east and north, respectively.

(a) Find the positions of the boats when $t = 0, 2, 4$ and 6 s.

(b) What happens when $t = 6$?

(c) Plot the paths of the two boats.

Solution

(a) The table below shows how the positions are calculated, by substituting the required values of t.

t	$\mathbf{r}_A$	$\mathbf{r}_B$
0	$2 \times 0\mathbf{i} + 4 \times 0\mathbf{j} = 0\mathbf{i} + 0\mathbf{j}$	$(8 \times 0 - 0^2)\mathbf{i} + 4 \times 0\mathbf{j} = 0\mathbf{i} + 0\mathbf{j}$
2	$2 \times 2\mathbf{i} + 4 \times 2\mathbf{j} = 4\mathbf{i} + 8\mathbf{j}$	$(8 \times 2 - 2^2)\mathbf{i} + 4 \times 2\mathbf{j} = 12\mathbf{i} + 8\mathbf{j}$
4	$2 \times 4\mathbf{i} + 4 \times 4\mathbf{j} = 8\mathbf{i} + 16\mathbf{j}$	$(8 \times 4 - 4^2)\mathbf{i} + 4 \times 4\mathbf{j} = 16\mathbf{i} + 16\mathbf{j}$
6	$2 \times 6\mathbf{i} + 4 \times 6\mathbf{j} = 12\mathbf{i} + 24\mathbf{j}$	$(8 \times 6 - 6^2)\mathbf{i} + 4 \times 6\mathbf{j} = 12\mathbf{i} + 24\mathbf{j}$

(b) When $t = 6$ the boats both have the same position at the same time and so will collide.

(c) The paths of the boats are shown below.

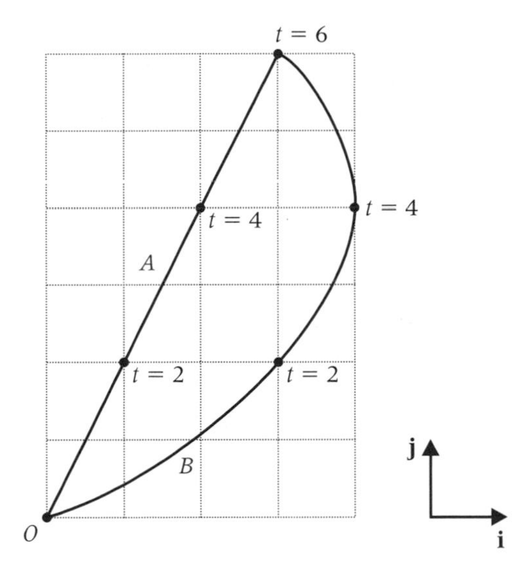

Worked example 3.3

A boat moves so that its position, in metres, at time t seconds is given by

$$\mathbf{r} = (450 - 5t)\mathbf{i} + (4t + 100)\mathbf{j}$$

where $\mathbf{i}$ and $\mathbf{j}$ are unit vectors that are directed east and north, respectively. A rock has position $250\mathbf{i} + 200\mathbf{j}$.

(a) Calculate the time when the boat is due north of the rock.

(b) Calculate the time when the boat is due east of the rock.

Solution

(a) When the boat is due north of the rock, its position will be $250\mathbf{i} + y\mathbf{j}$, where y is a constant. Equating the $\mathbf{i}$ components of the position vectors gives

$$250 = 450 - 5t$$

$$t = \frac{200}{5} = 40 \text{ s}$$

At this time the position of the boat is $250\mathbf{i} + 260\mathbf{j}$.

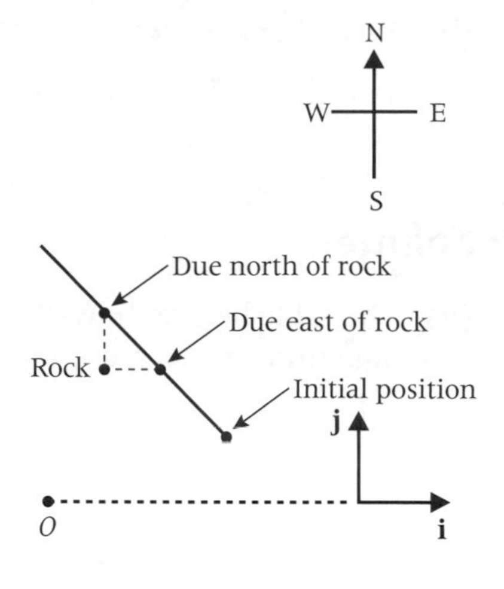

(b) When the boat is due east of the rock, its position will be $x\mathbf{i} + 200\mathbf{j}$, where x is a constant. Equating the $\mathbf{j}$ components of the position vectors gives

$$200 = 100 + 4t$$

$$t = \frac{100}{4} = 25 \text{ s}$$

At this time the position of the boat will be $325\mathbf{i} + 200\mathbf{j}$.

EXERCISE 3A

1 A golf ball is hit so that its position, in metres, at time t seconds is given by

$$\mathbf{r} = 30t\mathbf{i} + (25t - 4.9t^2)\mathbf{j}$$

where $\mathbf{i}$ and $\mathbf{j}$ are horizontal and vertical unit vectors, respectively.

(a) Find the position of the ball when $t = 0, 1, 2, 3, 4, 5$ and 6 s.

(b) Plot the path of the ball.

(c) From your plot estimate the horizontal distance travelled by the ball when it hits the ground.

2 A ball moves so that its position vector, in metres, at time t seconds is given by:

$$\mathbf{r} = 5t\mathbf{i} + (1 + 4t - 5t^2)\mathbf{j}$$

where $\mathbf{i}$ and $\mathbf{j}$ are horizontal and vertical unit vectors, respectively.

(a) Find the position of the ball when $t = 0, 0.2, 0.4, 0.6$ and 1 s.

(b) Use your answers to **(a)** to sketch the path of the ball.

3 Two children, A and B, run so that their position vectors in metres at time t seconds are given by:

$$\mathbf{r}_A = t\mathbf{i} + t\mathbf{j} \quad \text{and} \quad \mathbf{r}_B = (4 + 4t - t^2)\mathbf{i} + t\mathbf{j}$$

where **i** and **j** are unit vectors directed east and north, respectively.

Plot the paths of the two children for $0 \leq t \leq 4$. What happens when $t = 4$?

4 A boat moves so that its position, in metres, at time t seconds is given by

$$\mathbf{r} = (2t + 6)\mathbf{i} + (3t - 9)\mathbf{j}$$

where **i** and **j** are unit vectors that are directed east and north, respectively. A lighthouse has position $186\mathbf{i} + 281\mathbf{j}$.

(a) Find the position of the boat, when $t = 0$, 40, 80 and 120 s.

(b) Plot the path of the boat.

(c) From your plot of the path find the shortest distance between the boat and the lighthouse.

(d) Calculate the time when the boat is due south of the lighthouse and the time when the boat is due east of the lighthouse.

5 A bullet is fired from a rifle, so that its position, in metres, at time t seconds is given by

$$\mathbf{r} = 180t\mathbf{i} + (1.225 - 4.9t^2)\mathbf{j}$$

where **i** and **j** are horizontal and vertical unit vectors, respectively, and the origin is at ground level.

(a) Find the initial height of the bullet.

(b) Find the time when the bullet hits the ground.

(c) Find the horizontal distance travelled by the bullet.

6 A boomerang is thrown. As it moves its position, in metres, at time t seconds is modelled by

$$\mathbf{r} = (10t - t^2)\mathbf{i} + 4t\mathbf{j} + (2 + t - t^2)\mathbf{k}$$

where **i** and **j** are perpendicular unit vectors and **k** is a vertical unit vector. The origin is at ground level.

Find the position of the boomerang when it hits the ground.

3.3 Expressing quantities as vectors

Not all quantities such as positions, velocities or accelerations will be expressed in the form $a\mathbf{i} + b\mathbf{j}$. For example, a position may be expressed in terms of a distance and a bearing. In this section we see how to express these quantities in this form.

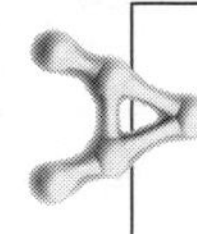

Consider a point that is at a distance d from the origin as shown in the diagram. If the angle between the unit vector $\mathbf{i}$ and the vector representing the position is θ, then we can write

$$\mathbf{r} = d\cos\theta\,\mathbf{i} + d\sin\theta\,\mathbf{j}$$

If $\mathbf{i}$ and $\mathbf{j}$ are horizontal and vertical, respectively, we would say that the horizontal component of the position vector is $d\cos\theta$ and that the vertical component is $d\sin\theta$

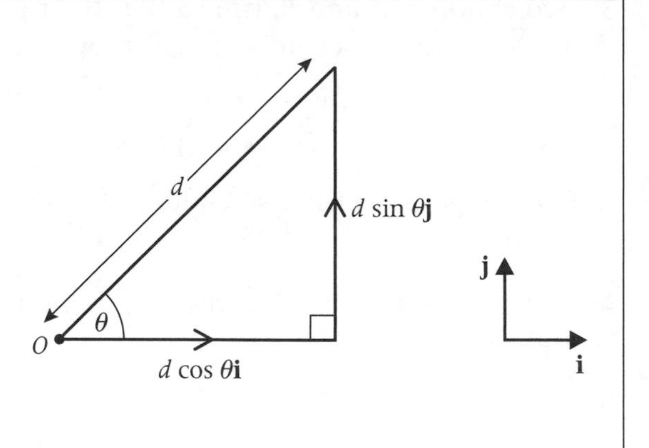

The following examples will illustrate how to write quantities in the form $a\mathbf{i} + b\mathbf{j}$.

Worked example 3.4

A ship starts at the origin and travels 200 km on a bearing of 140°. Express the position of the ship in the form $a\mathbf{i} + b\mathbf{j}$, where $\mathbf{i}$ and $\mathbf{j}$ are unit vectors that are directed east and north, respectively.

Solution

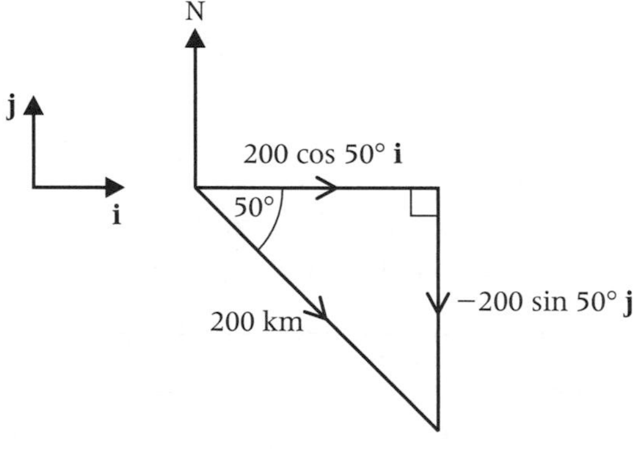

The diagram shows the position of the ship and the unit vectors.

In this case we can write:

$$\mathbf{r} = 200\cos 50°\,\mathbf{i} - 200\sin 50°\,\mathbf{j}$$

Worked example 3.5

A model aeroplane is travelling at $15\ \mathrm{m\,s^{-1}}$ on a bearing of 250°. Express the velocity of the aeroplane in the form $a\mathbf{i} + b\mathbf{j}$, where $\mathbf{i}$ and $\mathbf{j}$ are unit vectors that are directed east and north, respectively.

Solution

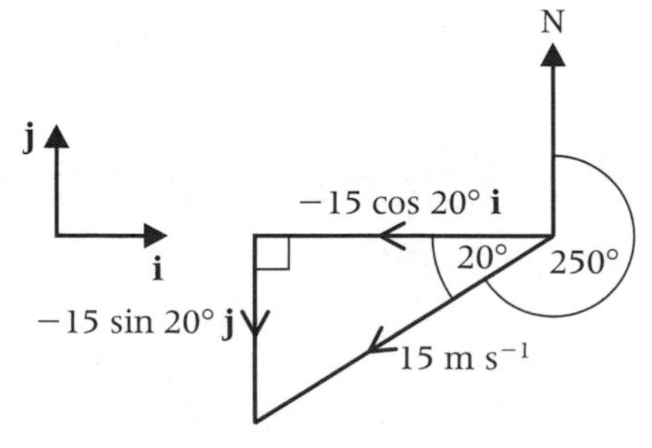

The diagram shows the velocity of the aeroplane and the unit vectors $\mathbf{i}$ and $\mathbf{j}$. In this case the velocity of the aeroplane can be expressed as

$$\mathbf{v} = -15\cos 20°\,\mathbf{i} - 15\sin 20°\,\mathbf{j}$$

Worked example 3.6

The velocity of a bird is $3\mathbf{i} + 4\mathbf{j}$, where $\mathbf{i}$ and $\mathbf{j}$ are unit vectors directed east and north, respectively. Find the speed of the bird and the direction in which it is heading.

Solution

The diagram shows the velocity of the bird and the unit vectors. The speed of the bird is given by the magnitude or length of the velocity vector. This can be calculated using Pythagoras' theorem.

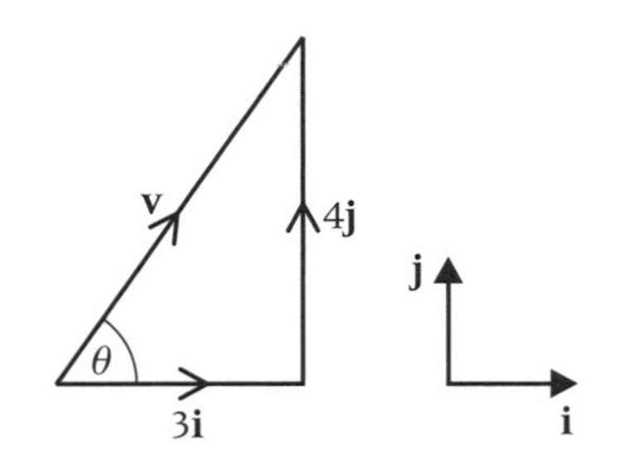

$$v = \sqrt{3^2 + 4^2}$$
$$= \sqrt{25}$$
$$= 5 \text{ m s}^{-1}$$

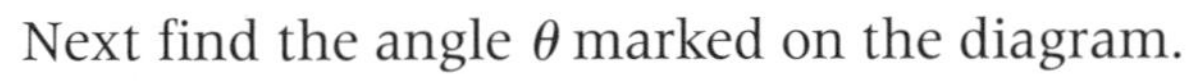

Next find the angle θ marked on the diagram.

$$\tan\theta = \frac{4}{3}$$
$$\theta = 53.1°$$

The best way to describe the direction of the velocity of the bird is by using the bearing of the direction that it is heading in. This is given by

$$90 - 53.1 = 036.9°$$

Note. The acceleration due to gravity is often expressed as $-g\mathbf{j}$, because it has magnitude g m s^{-2} and acts vertically downwards.

EXERCISE 3B

1 The unit vectors $\mathbf{i}$ and $\mathbf{j}$ are directed east and north, respectively. The positions below are given in terms of a bearing and a distance. Express each position in the form $a\mathbf{i} + b\mathbf{j}$.

(a) 45 m on a bearing of 080°.

(b) 105 m on a bearing of 060°.

(c) 21 m on a bearing of 340°.

(d) 62 m on a bearing of 260°.

(e) 290 m on a bearing of 162°.

2 A boat sails south west at 6 m s^{-1}. Find the velocity of the ship in the form $a\mathbf{i} + b\mathbf{j}$, where $\mathbf{i}$ and $\mathbf{j}$ are unit vectors directed east and north, respectively.

3 A ship travels at a speed of 5 m s^{-1}. Express its velocity in terms of the unit vectors **i** and **j**, that are directed east and north, respectively, if the ship is sailing:

(a) due east,

(b) due south,

(c) due west,

(d) south east,

(e) north west.

4 A ball is thrown so that its initial velocity is 8 m s^{-1} at an angle of 50° above the horizontal. The unit vectors **i** and **j** are horizontal and vertical, respectively. Express the initial velocity of the ball in terms of the unit vectors **i** and **j**.

5 For each velocity listed below, find its magnitude and direction. The unit vectors **i** and **j** are directed east and north, respectively. Give the directions as the bearing along which the velocity is directed.

(a) $(4\mathbf{i} + 7\mathbf{j}) \text{ m s}^{-1}$,

(b) $(5\mathbf{i} - 6\mathbf{j}) \text{ m s}^{-1}$,

(c) $(-8\mathbf{i} - 9\mathbf{j}) \text{ m s}^{-1}$,

(d) $(-12\mathbf{i} + 8\mathbf{j}) \text{ m s}^{-1}$.

6 An object moves along a straight line from the point with position vector $(5\mathbf{i} + 6\mathbf{j})$ m to the point with position vector $(8\mathbf{i} - 2\mathbf{j})$ m, where the unit vectors **i** and **j** are directed east and north, respectively.

(a) Find the distance travelled by the object.

(b) Find the bearing along which the object was travelling.

7 The acceleration of a particle is $(-3\mathbf{i} + 2\mathbf{j}) \text{ m s}^{-2}$, where the unit vectors **i** and **j** are directed north and east, respectively. Find the magnitude of the acceleration. Calculate the bearing along which the acceleration is directed.

3.4 Constant acceleration equations in two and three dimensions

The constant acceleration equations that were introduced and used in the last chapter can be extended into two and three dimensions as vector equations.

To make use of these vector equations the velocities, accelerations and positions must be in the form $a\mathbf{i} + b\mathbf{j}$ for two-dimensional motion, or $a\mathbf{i} + b\mathbf{j} + c\mathbf{k}$ for three dimensions.

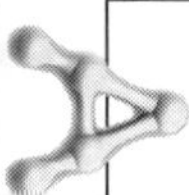

The constant acceleration equations become

$$\mathbf{r} = \mathbf{u}t + \frac{1}{2}\mathbf{a}t^2 \quad \text{or} \quad \mathbf{r} = \mathbf{u}t + \frac{1}{2}\mathbf{a}t^2 + \mathbf{r}_0$$

$$\mathbf{v} = \mathbf{u} + \mathbf{a}t$$

$$\mathbf{r} = \frac{1}{2}(\mathbf{u} + \mathbf{v})t \quad \text{or} \quad \mathbf{r} = \frac{1}{2}(\mathbf{u} + \mathbf{v})t + \mathbf{r}_0$$

where $\mathbf{r}$ is the position at time t, $\mathbf{u}$ is the initial velocity, $\mathbf{v}$ is the velocity at time t, $\mathbf{a}$ is the acceleration and $\mathbf{r}_0$ is the initial position.

You will notice the similarity between these equations and the constant acceleration equations that you used in the last chapter. At this stage we will not use a vector equation that is equivalent to $v^2 = u^2 + 2as$.

The following examples illustrate how these formulae can be applied.

Worked example 3.7

A ball is rolling on an inclined plane. The initial velocity of the ball is $4\mathbf{i}$ m s^{-1}, its acceleration is $-2\mathbf{j}$ m s^{-2} and its initial position is $(9\mathbf{i} + 4\mathbf{j})$ m, where $\mathbf{i}$ and $\mathbf{j}$ are perpendicular unit vectors that lie in the plane on which the ball is moving.

(a) Find the velocity of the ball after it has been moving for 3 s.

(b) Find the position of the ball after it has been moving for 5 s.

Solution

For this problem:

$$\mathbf{u} = 4\mathbf{i}$$

$$\mathbf{a} = -2\mathbf{j}$$

$$\mathbf{r}_0 = 9\mathbf{i} + 4\mathbf{j}$$

(a) Using the formula for the velocity with the vectors above and $t = 3$ gives:

$$\begin{aligned} \mathbf{v} &= \mathbf{u} + \mathbf{a}t \\ &= 4\mathbf{i} + (-2\mathbf{j}) \times 3 \\ &= 4\mathbf{i} - 6\mathbf{j} \end{aligned}$$

(b) Using the formula for the position with the vectors listed above and $t = 5$ gives:

$$\begin{aligned}\mathbf{r} &= \mathbf{u}t + \frac{1}{2}\mathbf{a}t^2 + \mathbf{r}_0 \\ &= 4\mathbf{i} \times 5 + \frac{1}{2}(-2\mathbf{j}) \times 5^2 + 9\mathbf{i} + 4\mathbf{j} \\ &= 20\mathbf{i} - 25\mathbf{j} + 9\mathbf{i} + 4\mathbf{j} \\ &= 29\mathbf{i} - 21\mathbf{j}\end{aligned}$$

Worked example 3.8

A boat has initial velocity $(5\mathbf{i} + 3\mathbf{j})$ m s^{-1}. It accelerates for 4 s. Its velocity is then $(2\mathbf{i} - 4\mathbf{j})$ m s^{-1}. The boat then stops accelerating and travels with this velocity for a further 20 s. The initial position of the boat is $(24\mathbf{i} + 32\mathbf{j})$ m. The unit vectors **i** and **j** are directed east and north, respectively.

(a) Find the acceleration of the boat.

(b) Find the position of the boat after 4 s.

(c) Find the final position of the boat.

Solution

(a) The acceleration can be found using the equation $\mathbf{v} = \mathbf{u} + \mathbf{a}t$. Using $\mathbf{u} = 5\mathbf{i} + 3\mathbf{j}$, $\mathbf{v} = 2\mathbf{i} - 4\mathbf{j}$ and $t = 4$ gives:

$$\begin{aligned}2\mathbf{i} - 4\mathbf{j} &= 5\mathbf{i} + 3\mathbf{j} + 4\mathbf{a} \\ 4\mathbf{a} &= -3\mathbf{i} - 7\mathbf{j} \\ \mathbf{a} &= -\frac{3}{4}\mathbf{i} - \frac{7}{4}\mathbf{j}\end{aligned}$$

(b) The position after 4 s can be found using the constant acceleration equation $\mathbf{r} = \frac{1}{2}(\mathbf{u} + \mathbf{v})t + \mathbf{r}_0$, with $\mathbf{u} = 5\mathbf{i} + 3\mathbf{j}$, $\mathbf{v} = 2\mathbf{i} - 4\mathbf{j}$, $\mathbf{r}_0 = 24\mathbf{i} + 32\mathbf{j}$ and $t = 4$.

$$\begin{aligned}\mathbf{r} &= \frac{1}{2}(\mathbf{u} + \mathbf{v})\mathbf{t} + \mathbf{r}_0 \\ &= \frac{1}{2}(5\mathbf{i} + 3\mathbf{j} + 2\mathbf{i} - 4\mathbf{j}) \times 4 + 24\mathbf{i} + 32\mathbf{j} \\ &= 14\mathbf{i} - 2\mathbf{j} + 24\mathbf{i} + 32\mathbf{j} \\ &= 38\mathbf{i} + 30\mathbf{j}\end{aligned}$$

(c) After 4 s the boat moves with a constant velocity for a further 20 s. As the acceleration is zero the equation $\mathbf{r} = \mathbf{u}t + \frac{1}{2}\mathbf{a}t^2 + \mathbf{r}_0$ reduces to $\mathbf{r} = \mathbf{u}t + \mathbf{r}_0$. This can be applied by considering only the motion after the boat stops accelerating. Using $\mathbf{u} = 2\mathbf{i} - 4\mathbf{j}$, $\mathbf{r}_0 = 38\mathbf{i} + 30\mathbf{j}$ and $t = 20$ gives

$$\begin{aligned}\mathbf{r} &= \mathbf{u}t + \mathbf{r}_0\\ &= (2\mathbf{i} - 4\mathbf{j}) \times 20 + 38\mathbf{i} + 30\mathbf{j}\\ &= 78\mathbf{i} - 50\mathbf{j}\end{aligned}$$

Worked example 3.9

An aeroplane has a constant velocity of $120\mathbf{i}$ m s^{-1}, as it is moving along a runway. It then experiences an acceleration of $(2\mathbf{i} + 5\mathbf{j})$ m s^{-2} for the first 20 s of its flight. The unit vectors $\mathbf{i}$ and $\mathbf{j}$ are directed horizontally and vertically, respectively. Assume that the aeroplane is at the origin when it begins to accelerate.

(a) Find an expression for the position of the aeroplane at time t s, after it starts to accelerate.

(b) Find the speed of the aeroplane when it is at a height of 250 m.

Solution

(a) As the aeroplane is initially at the origin we can use the constant acceleration equation $\mathbf{r} = \mathbf{u}t + \frac{1}{2}\mathbf{a}t^2$, with $\mathbf{u} = 120\mathbf{i}$ and $\mathbf{a} = 2\mathbf{i} + 5\mathbf{j}$.

$$\begin{aligned}\mathbf{r} &= \mathbf{u}t + \frac{1}{2}\mathbf{a}t^2\\ &= 120t\mathbf{i} + \frac{1}{2}(2\mathbf{i} + 5\mathbf{j})t^2\\ &= (120t + t^2)\mathbf{i} + \frac{5}{2}t^2\mathbf{j}\end{aligned}$$

(b) The height of the aeroplane is given by the vertical or $\mathbf{j}$ component of the position vector. When the height of the aeroplane is 250 m we have:

$$\begin{aligned}\frac{5}{2}t^2 &= 250\\ t^2 &= 100\\ t &= 10 \text{ s}\end{aligned}$$

The velocity can now be found using this value for t.

$$\mathbf{v} = \mathbf{u} + \mathbf{a}t$$
$$= 120\mathbf{i} + (2\mathbf{i} + 5\mathbf{j}) \times 10$$
$$= 140\mathbf{i} + 50\mathbf{j}$$

The speed can be found as it will be the magnitude of the velocity.

$$v = \sqrt{140^2 + 50^2}$$
$$= \sqrt{22100}$$
$$= 149 \text{ m s}^{-1} \text{ (to 3 sf)}$$

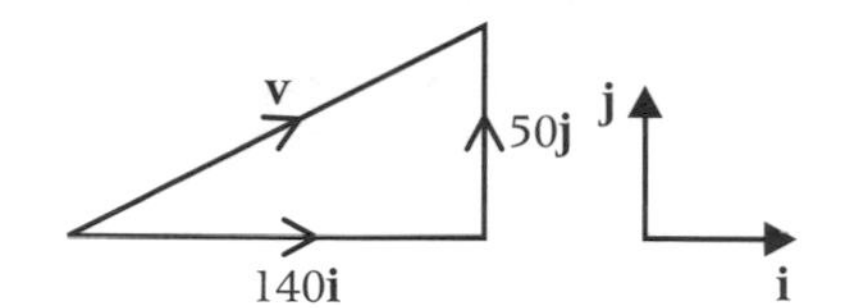

EXERCISE 3C

1 During a 10 s period the velocity of a boat changes from $(4\mathbf{i} + 2\mathbf{j})$ m s^{-1} to $(\mathbf{i} - 3\mathbf{j})$ m s^{-1}, where $\mathbf{i}$ and $\mathbf{j}$ are perpendicular unit vectors. Find the acceleration of the boat during this time, assuming that it is constant.

2 The acceleration of a motor boat is $(0.6\mathbf{i} + 0.8\mathbf{j})$ m s^{-2}. Its initial velocity is $3\mathbf{i}$ m s^{-1} and its initial position is $(20\mathbf{i} + 5\mathbf{j})$ m. The unit vectors, $\mathbf{i}$ and $\mathbf{j}$ are directed east and north, respectively.

(a) Find the velocity of the boat when $t = 3$ s.

(b) Find the position of the boat when $t = 3$ s.

3 An object has initial velocity $(3\mathbf{i} - 5\mathbf{j})$ m s^{-1} and an acceleration of $(\mathbf{i} + \mathbf{j})$ m s^{-2}, where $\mathbf{i}$ and $\mathbf{j}$ are perpendicular unit vectors. If it starts at the origin find the position and velocity at time t s.

4 The acceleration of a body is $(6\mathbf{i} + 8\mathbf{j})$ m s^{-2}. If the body starts at rest at $0\mathbf{i} + 0\mathbf{j}$ and accelerates for 6 s find the velocity and position of the body after 6 s. The unit vectors $\mathbf{i}$ and $\mathbf{j}$ are perpendicular.

5 A ball has initial position $2\mathbf{j}$ m, initial velocity $(4\mathbf{i} + 9\mathbf{j})$ m s^{-1} and acceleration $-10\mathbf{j}$ m s^{-2}, where $\mathbf{i}$ and $\mathbf{j}$ are horizontal and vertical unit vectors, respectively. Find the position of the ball at time t s and its position when it hits the ground, that is when the vertical component of its position is zero.

6 A snooker ball is struck so that it moves with the constant velocity shown in the diagram. It starts at the point with position vector $(0.2\mathbf{i} + 0.1\mathbf{j})$ m relative to the origin O. Find an expression for the position of the ball at time t, and find where it first hits a cushion. The unit vectors $\mathbf{i}$ and $\mathbf{j}$ are directed as shown on the diagram.

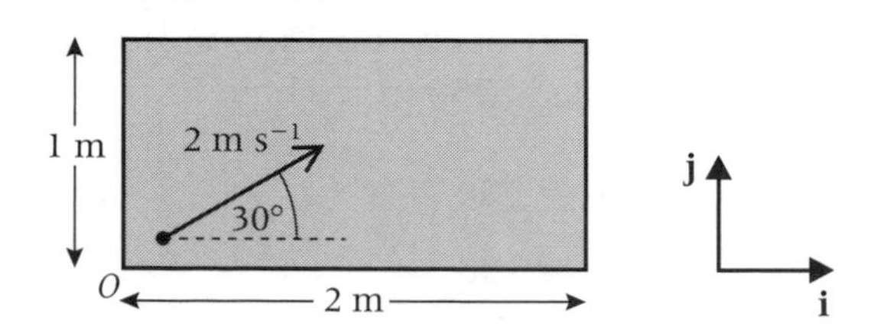

7 A ball is launched from a point with position $(0\mathbf{i} + 1.9\mathbf{j})$ m and velocity $(7\mathbf{i} + 32\mathbf{j})$ m s^{-1}. The ball experiences an acceleration of $(-9.8\mathbf{j})$ m s^{-2} while in the air. The unit vectors $\mathbf{i}$ and $\mathbf{j}$ are horizontal and vertical, respectively.

(a) Show that the position of the ball, in metres, at time t seconds is given by:

$$\mathbf{r} = 7t\mathbf{i} + (1.9 + 32t - 4.9t^2)\mathbf{j}$$

(b) Find how long the ball is in the air and where it lands.

(c) Find the maximum height reached by the ball.

(d) Find the speed of the ball when it hits the ground.

8 A golf ball is hit from ground level so that its initial velocity is $(20\mathbf{i} + 30\mathbf{j})$ m s^{-1} and its acceleration is $-10\mathbf{j}$ m s^{-2}, where $\mathbf{i}$ and $\mathbf{j}$ are horizontal and vertical unit vectors, respectively. Assume that the initial position of the ball is $0\mathbf{i} + 0\mathbf{j}$.

(a) Find the time that the ball is in the air, the position of the point where it hits the ground and its speed at this time.

(b) Find the time for which the height of the ball is greater than 25 m.

9 The unit vectors $\mathbf{i}$ and $\mathbf{j}$, are directed north and east, respectively. A boat has initial velocity $(4\mathbf{i} + 6\mathbf{j})$ m s^{-1} and an initial position of $(80\mathbf{i} + 20\mathbf{j})$ m. It experiences an acceleration of $(-0.02\mathbf{i} - 0.04\mathbf{j})$ m s^{-2} for a period of 4 minutes.

(a) Find an expression for the velocity and position of the particle at time t s.

(b) After 4 minutes the boat stops accelerating and continues with a constant velocity for a further minute before hitting the bank and stopping. Find the position of the point where the boat hits the bank.

(c) Find the velocity of the boat when its position is $476\mathbf{i} + 452\mathbf{j}$.

10 A model helicopter has initial position $80\mathbf{j}$ m and flies with a constant velocity of $(6\mathbf{i} - 3\mathbf{j})$ m s^{-1}. A model aeroplane flies at the same height as the helicopter. Its initial position is $10\mathbf{j}$ m, its initial velocity is $5\mathbf{i}$ m s^{-1} and its acceleration is $(0.1\mathbf{i} + 0.05\mathbf{j})$ m s^{-2}. The unit vectors $\mathbf{i}$ and $\mathbf{j}$ are directed east and north, respectively.

(a) Find expressions for the position vectors of the helicopter and the aeroplane at time t s.

(b) The helicopter and the aeroplane collide. Find the time when this collision takes place and the position of the collision.

11 The unit vectors **i** and **j**, are directed east and north, respectively. A boat has initial velocity $(2\mathbf{i} + 3\mathbf{j})$ m s^{-1} and an initial position of $(40\mathbf{i} + 20\mathbf{j})$ m with respect to an origin O. It experiences an acceleration of $(-0.06\mathbf{i} - 0.04\mathbf{j})$ m s^{-2}. Model the boat as a particle.

(a) Find expressions for the velocity and position, with respect to O, of the boat at time t s.

(b) The boat hits a sandbank when its position is $52\mathbf{i} + 128\mathbf{j}$. Find the value of t when this happens. [A]

12 A model assumes that an aeroplane has an initial velocity of $200\mathbf{i}$ m s^{-1} and experiences an acceleration of $(-0.5\mathbf{i} - 0.05\mathbf{j})$ m s^{-2} in preparation for landing. The initial position of the aeroplane was $(-50\,000\mathbf{i} + 4000\mathbf{j})$ m with respect to an origin O. The unit vectors **i** and **j** are horizontal and vertical, respectively. After accelerating for 200 s the velocity of the aeroplane is assumed to remain constant until it lands. The aeroplane lands when the vertical component of its position vector is zero.

(a) Find:

(i) the time it takes for the aeroplane to move from its initial position to the point where it first touches the ground,

(ii) the position of the aeroplane when it first touches the ground,

(iii) the speed of the aeroplane when it first touches the ground.

(b) Comment on the assumption that the velocity of the aeroplane is constant during the first stage of its flight. [A]

Key point summary

Formulae to learn

Constant acceleration equations for two or three dimensions

$$\mathbf{r} = \mathbf{u}t + \frac{1}{2}\mathbf{a}t^2 \quad \text{or} \quad \mathbf{r} = \mathbf{u}t + \frac{1}{2}\mathbf{a}t^2 + \mathbf{r}_0$$

$$\mathbf{v} = \mathbf{u} + \mathbf{a}t$$

$$\mathbf{r} = \frac{1}{2}(\mathbf{u} + \mathbf{v})t \quad \text{or} \quad \mathbf{r} = \frac{1}{2}(\mathbf{u} + \mathbf{v})t + \mathbf{r}_0$$

1 A position vector enables paths to be plotted and interpreted. *p29*

2 Positions, velocities and accelerations can be expressed in the form $x\mathbf{i} + y\mathbf{j} + z\mathbf{k}$ or $x\mathbf{i} + z\mathbf{j}$. *p29*

3 Magnitudes and directions of vectors can be found using trigonometric functions. *p34*

4 Constant acceleration equations can be used in two or three dimensions. *p37*

Test yourself	What to review
1 A ball moves so that its position, in metres, at time t seconds is given by $\mathbf{r} = 5t\mathbf{i} + (6 + 9.5t - 5t^2)\mathbf{j}$ where the unit vectors $\mathbf{i}$ and $\mathbf{j}$ are horizontal and vertical, respectively. The origin is at ground level. **(a)** Find the time when the ball hits the ground. **(b)** Plot the path of the ball. **(c)** Estimate the maximum height of the ball and the horizontal distance travelled by the ball from your plot.	*Section 3.2*
2 A ship travels at 3 m s^{-1} on a bearing of 235°. Express this velocity in the form $a\mathbf{i} + b\mathbf{j}$, where $\mathbf{i}$ and $\mathbf{j}$ are unit vectors that are directed east and north, respectively.	*Section 3.3*
3 The velocity of an aeroplane is $(80\mathbf{i} + 50\mathbf{j})$ m s^{-1}, where $\mathbf{i}$ and $\mathbf{j}$ are unit vectors that are directed east and north, respectively. Find the speed of the aeroplane and the direction in which it is heading.	*Section 3.3*
4 The unit vectors $\mathbf{i}$ and $\mathbf{j}$ are perpendicular and lie in a horizontal plane. A particle moves from the origin. Its initial velocity was $(4\mathbf{i} + 6\mathbf{j})$ m s^{-1} and after 20 s its velocity is $(24\mathbf{i} + 46\mathbf{j})$ m s^{-1}. The acceleration of the particle is constant. **(a)** Find the acceleration of the particle. **(b)** Find the distance of the particle from the origin after accelerating for 30 s.	*Section 3.4*

Test yourself ANSWERS

1 **(a)** 2.4 s, **(c)** 10.5 m, 12 m.

2 $-3\cos 35°\,\mathbf{i} - 3\sin 35°\,\mathbf{j}$.

3 94.3 m s^{-1}, 058.0°.

4 **(a)** $\mathbf{i} + 2\mathbf{j}$, **(b)** 1221 m.

CHAPTER 4

Forces

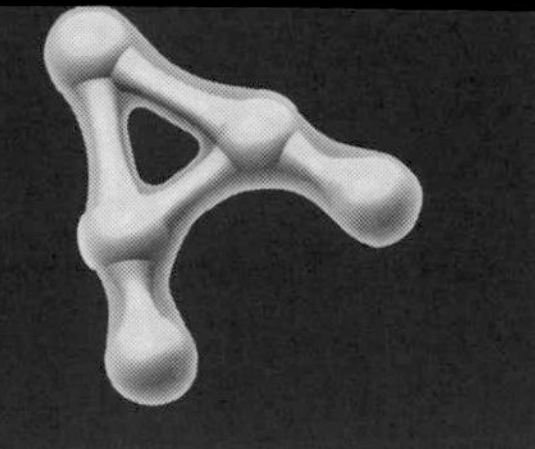

Learning objectives

After studying this chapter you should be able to:

- identify forces acting on a body
- draw force diagrams
- resolve forces into components
- find resultant forces
- write forces as vectors
- understand that the resultant force is zero when the forces are in equilibrium
- know that for equilibrium a body must be at rest or moving with a constant speed
- use the friction inequality.

4.1 Introduction

In the earlier chapters of this book we have considered how to describe motion, using terms like velocity, acceleration and position. In this chapter we will consider forces. Forces cause motion or can act to keep objects at rest. An understanding of forces and how they cause motion is essential to be able to predict how object in real life will move.

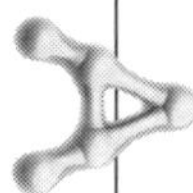

Force is a vector quantity – it has magnitude and direction. The unit of force is the newton, which is abbreviated to N.

We will look at the different types of force that can act on a body, and learn how to add them together to find their sum, which is called their resultant. We will investigate a particle in equilibrium. This is where the resultant force on the particle is zero, so that the particle remains at rest or moves with a constant velocity. Finally, we will be looking at frictional forces. We will be required to find unknowns such as forces, angles, and so on. In order to do this it is essential that we are able to draw a clear force diagram, which shows all the forces acting on the particle.

4.2 Types of force

Weight

Weight is a force, which is the effect of the earth's gravitational pull. If this force acted on its own on a body it would cause it to accelerate. This acceleration is approximately 9.8 m s^{-2} (denoted by g). The force, which causes this acceleration, is mg N, where m is the mass of the particle, in kg.

Note that any two objects that are allowed to fall, will have the same acceleration and so fall at the same rate, even though the weight forces acting on them are different.

As forces are vectors, we will represent them on diagrams by arrows, which show the direction of the force. We will indicate the size or magnitude of the force by writing this next to the arrow. The diagram shows a force of magnitude F N and the direction in which it acts. In a sketch the length of the arrow is not important, but if you solve problems by scale drawing, then the length of the vector or arrow must be proportional to the magnitude of the force.

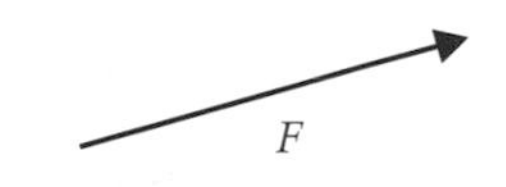

Particle on a plane

Suppose a particle with a weight of W N is at rest on a smooth horizontal plane. As the particle does not fall, the weight W is being balanced by an equal force acting upwards. This force is called the normal reaction, R, and acts perpendicular to the plane.

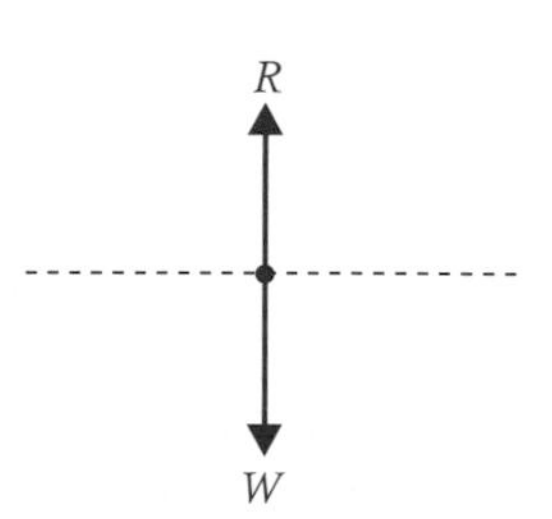

If we now place the particle on a rough horizontal plane, and pull the particle horizontally with a force of P N, then the roughness of the surface will oppose the motion. There will be a frictional force of F N, acting parallel to the plane, in the opposite direction. If the particle is at rest then $F = P$.

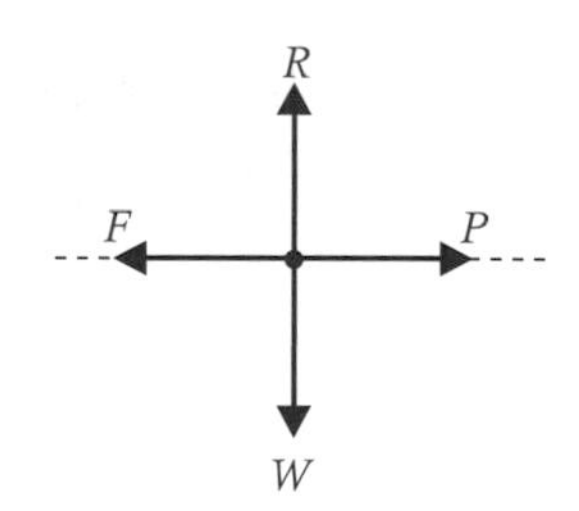

If we now consider the particle at rest on a rough plane, inclined at θ to the horizontal, then the frictional force will act up the slope, as shown in the diagram.

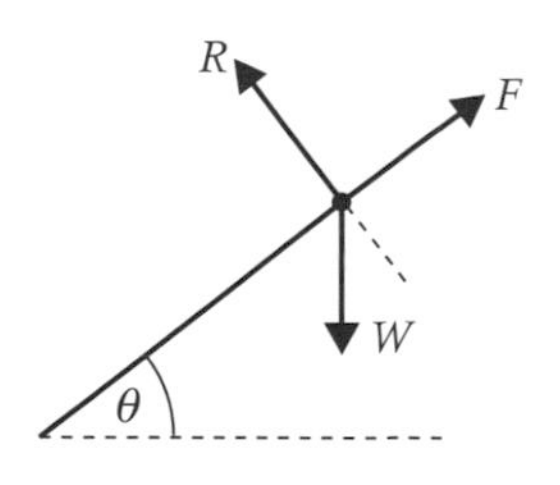

Strings and rods

A string will exert a force on an object if the string is taut, and this force, which is called a tension, will be directed along the string. For example, if a mass is suspended from a string, there will be a weight force acting downwards and a tension acting upwards.

Suppose a string is attached to a particle of weight W N, which rests on a smooth horizontal table. The only force, which the string can exert on the particle, will be in the same direction as the string itself. Furthermore, the string can only pull (and not push). Hence the tension, T, acts on the particle, as shown in the diagram.

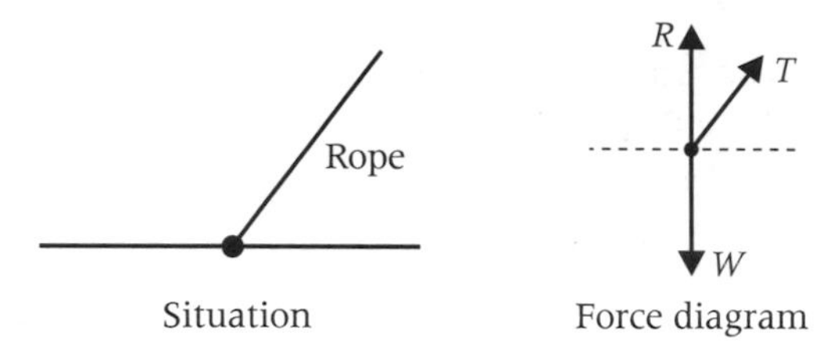

If a rod is attached to a particle instead of the string then the force of the rod on the particle could be in either of two directions, depending on whether it is pulling or pushing.

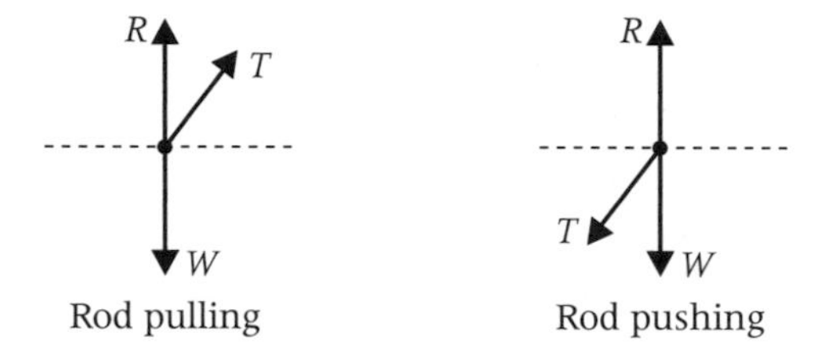

Here are some examples where the force diagram has been drawn to demonstrate particular situations.

1 A particle of weight W N sliding down a smooth plane inclined at α to the horizontal. Note that the term smooth means that there is no friction.

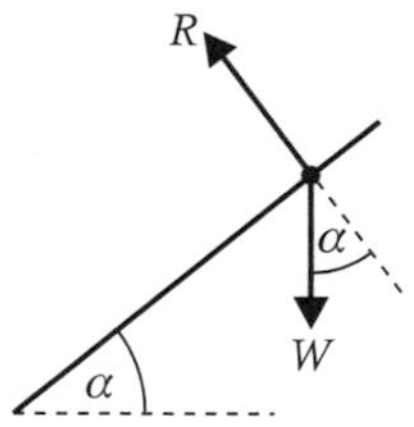

2 A particle of mass M kg at rest, suspended by a string.

3 A particle of mass M kg pulled across a rough horizontal plane by a string, inclined at 30° to the horizontal.

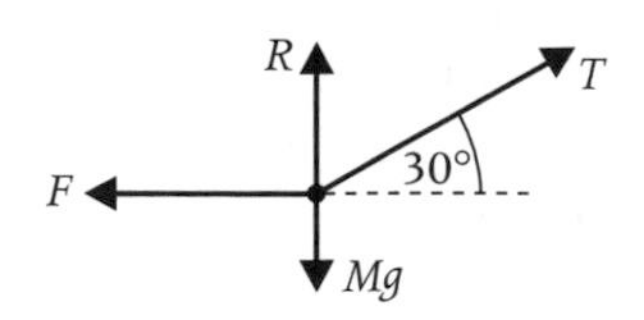

4 A and B are two points in the horizontal plane, a distance of 5 m apart. A particle, of weight W N, is attached by two strings, of length 3 m and 4 m, to the points A and B. The particle is at rest. The diagram shows the forces acting on the particle.

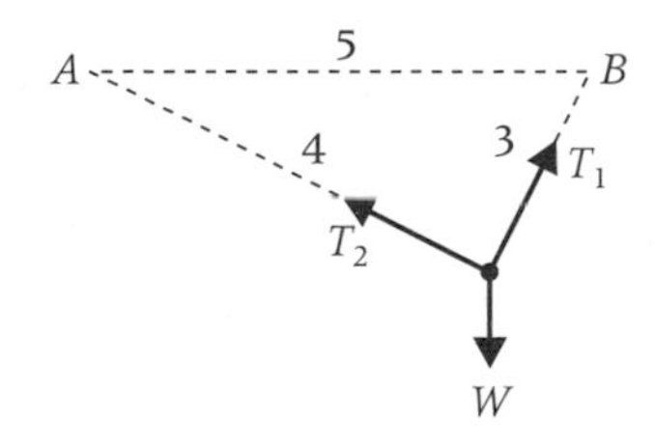

EXERCISE 4A

Draw force diagrams which show all the forces acting on the particle involved in each of the following situations:

1 A particle sliding down a rough plane inclined at α to the horizontal.

2 A particle of weight W N which is suspended from a fixed point by a string. The particle is held in equilibrium with the string at an angle θ to the vertical by a horizontal force P N.

3 A particle of mass m kg hangs in equilibrium supported by two light inextensible strings, inclined at 40° and 50° to the vertical.

4 A particle pushed up a rough plane, inclined at α to the horizontal, by a light horizontal rod, that is itself at an angle β to the slope.

5 A particle of weight W N is attached to one end of a light rod. The other end of the rod is fixed. A horizontal force P pulls the particle sideways, so that the rod makes an angle of 20° with the vertical.

4.3 Resultant forces

Adding two forces

Consider two forces, of magnitude F_1 and F_2, which act upon a particle. If we place these forces end to end, it can be seen that they have the same effect as a single force, of magnitude R. This force is known as the resultant force.

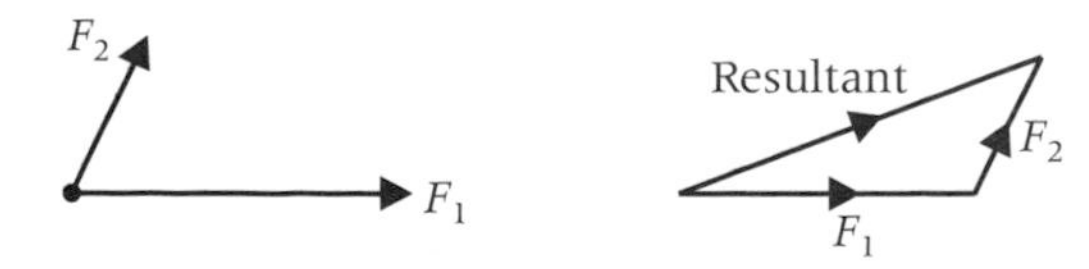

The resultant force will form the third side in a triangle of forces.

Worked example 4.1

Two forces of magnitudes 6 N and 5 N, act on a particle. The angle between the forces is 40°. Find the magnitude and direction of the resultant force.

Solution

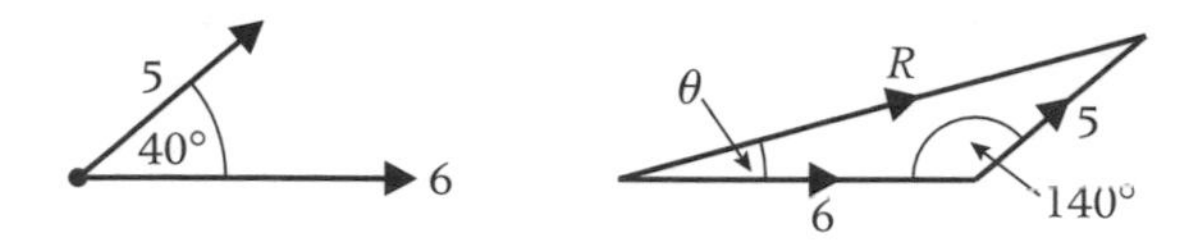

Using the cosine rule in the triangle of forces:

$$R^2 = 6^2 + 5^2 - 2 \times 6 \times 5 \times \cos 140°$$

$$R = 10.3 \text{ N}$$

Applying the sine rule:

$$\frac{\sin \theta}{5} = \frac{\sin 140}{R}$$

$$\theta = 18.1°$$

So the resultant force has magnitude 10.3 N and is at 18.1° to the 6 N force.

Worked example 4.2

Forces, of magnitude 14 N and 8 N, act on a particle. The resultant force has magnitude 17 N. Find the angle between the forces.

Solution

The required angle is θ, as shown. However, it is easier to first find the angle marked x, in the triangle of forces.

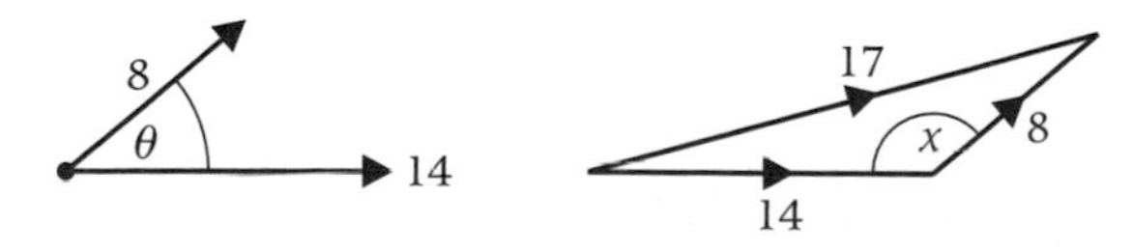

Using the cosine rule:

$$\cos x = \frac{14^2 + 8^2 - 17^2}{2 \times 14 \times 8}$$

$$x = 97.4°$$

The angle, θ, between the forces is:

$$180 - 97.4 = 82.6°$$

EXERCISE 4B

1 Find the magnitude of the resultant of the forces, of magnitude F_1 and F_2, if θ is the angle between them. Also find the angle between the resultant and the force of magnitude F_1.

(a) $F_1 = 4$ N, $F_2 = 3$ N, $\theta = 90°$,

(b) $F_1 = 6$ N, $F_2 = 10$ N, $\theta = 60°$,

(c) $F_1 = 7$ N, $F_2 = 9$ N, $\theta = 75°$,

(d) $F_1 = 8$ N, $F_2 = 12$ N, $\theta = 170°$,

(e) $F_1 = 10$ N, $F_2 = 11$ N, $\theta = 160°$.

2 Forces, of magnitude F_1 and F_2, act on a particle. The resultant of the forces has magnitude R. Find the angle between the two forces acting on the particle in the following cases:

(a) $F_1 = 60$ N, $F_2 = 80$ N, $R = 100$ N,

(b) $F_1 = 11$ N, $F_2 = 17$ N, $R = 20$ N,

(c) $F_1 = 7$ N, $F_2 = 10$ N, $R = 5$ N,

(d) $F_1 = 8$ N, $F_2 = 7$ N, $R = 3$ N.

3 Forces of magnitude 6 N and 5 N act on a particle. What are the greatest and least values of the magnitude of the resultant of these forces?

4 The resultant of two forces, of magnitude F_1 and F_2, has magnitude 20 N. If $F_1 = 8$ N and the angle between the two forces is 70°, find F_2.

5 The resultant of two forces, of magnitude P and Q has magnitude 15 N. If $Q = 6$ N and the angle between the two forces is 60°, find P.

6 The resultant of two forces, of magnitude F_1 and F_2 has magnitude 3 N. If $F_1 = 7$ N and $F_2 = 5$ N, calculate the angle between the two forces.

7 The resultant of two forces, of magnitude F_1 and F_2, has magnitude 12 N, and acts at an angle of 30° to F_1. If $F_1 = 6$ N find the magnitude of F_2 and direction of F_2.

4.4 Adding any number of forces

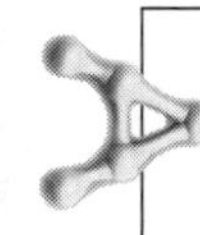

Three forces can be added by placing one force on the end of the other. The resultant of the forces is represented by the vector that can be drawn along the fourth side of the resulting quadrilateral, as shown below. The magnitude of the resultant, R, is given by the length of the fourth side.

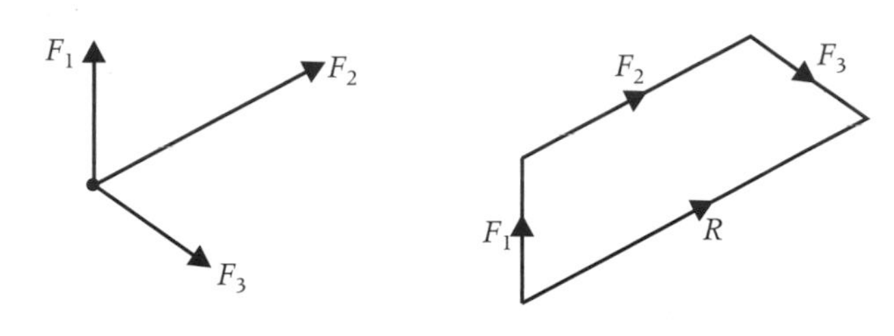

The quadrilateral can be divided into two triangles. The sine and cosine rules can then be used in the triangles to find the magnitude and direction of the resultant force.

Worked example 4.3

Find the magnitude of the resultant of the following set of forces:

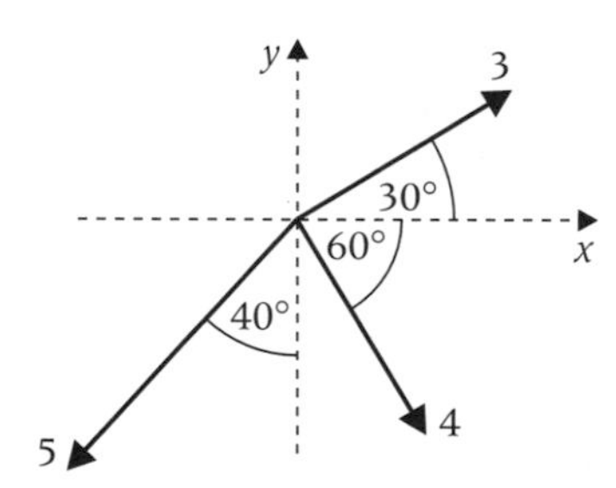

Solution

The diagram shows the three forces placed end to end to form a quadrilateral. The magnitude of the resultant force is R.

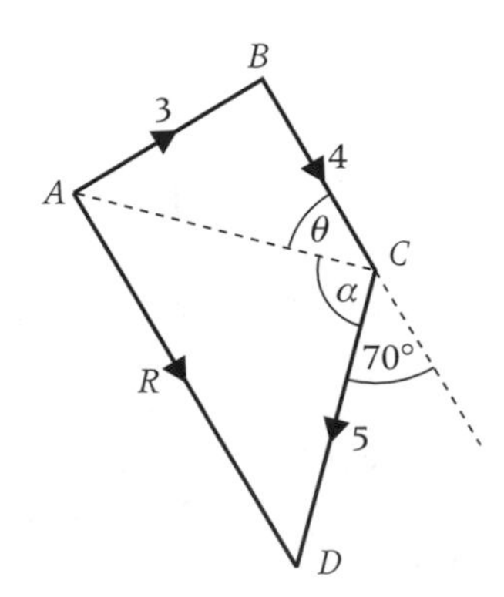

The angle between the 3 N force and the 4 N force is 90°. Therefore Pythagoras' theorem can be used to find AC in triangle ABC.

$$AC^2 = 3^2 + 4^2,$$

$$AC = 5.$$

Also $\tan \theta = \frac{3}{4}$

$$\theta = 36.9°.$$

$$\alpha = 180 - \theta - 70 = 73.1°$$

Using the cosine rule in triangle ACD gives:

$$R^2 = 5^2 + 5^2 - 2 \times 5 \times 5 \cos 73.1°$$

$$R = 5.96\text{ N}$$

It is also possible to show that the angle between R and the x-axis is 76.6°.

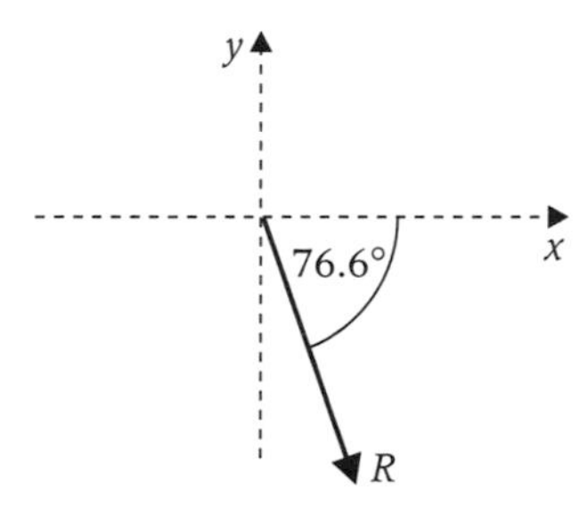

Using forces given in component form

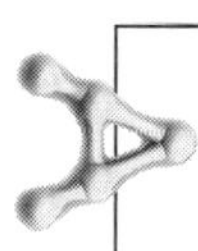

The approach of Worked example 4.3 is a slow and complicated process in general. The resultant of a set of forces can be found more quickly when the forces are given in component form.

In the same way that velocities and other vectors were expressed in the form $a\mathbf{i} + b\mathbf{j}$ in an earlier chapter, forces can be expressed in the same way. For example a force that has magnitude 40 N and acts at 30° above the horizontal could be expressed as:

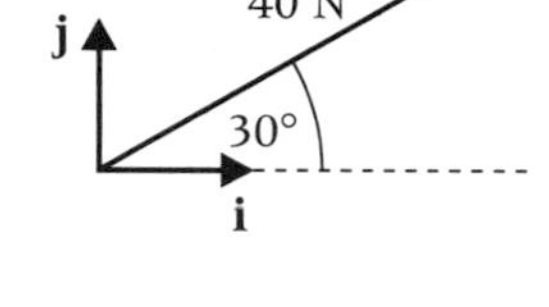

$$40 \cos 30°\,\mathbf{i} + 40 \sin 30°\,\mathbf{j}$$

First, we will concentrate on working with forces that are already expressed in vector form.

Worked example 4.4

Find the magnitude of the resultant of the following sets of forces:

$$\mathbf{F}_1 = (2\mathbf{i} + 3\mathbf{j} + \mathbf{k})\text{ N},\ \mathbf{F}_2 = (\mathbf{i} - 2\mathbf{j} + 4\mathbf{k})\text{ N},\ \mathbf{F}_3 = (2\mathbf{j} - 6\mathbf{k})\text{ N}.$$

Solution

First find the resultant in terms of **i** and **j**.

$$\mathbf{R} = \mathbf{F}_1 + \mathbf{F}_2 + \mathbf{F}_3$$

$$= 2\mathbf{i} + 3\mathbf{j} + \mathbf{k} + \mathbf{i} - 2\mathbf{j} + 4\mathbf{k} + 2\mathbf{j} - 6\mathbf{k}$$

$$= 3\mathbf{i} + 3\mathbf{j} - \mathbf{k}$$

Then find the magnitude of the force.

$$R = \sqrt{3^2 + 3^3 + (-1)^2}$$

$$= \sqrt{19} = 4.36\text{ N (to 3 sf)}$$

Worked example 4.5

The forces $(5\mathbf{i} + 12\mathbf{j})$ N and $(2\mathbf{i} + 4\mathbf{j})$ N act at a point.

(a) Find the magnitude of the resultant force.

(b) Find the angle between the resultant force and the unit vector **i**.

Solution

(a) Resultant $= (5\mathbf{i} + 12\mathbf{j}) + (2\mathbf{i} + 4\mathbf{j})$

$= 7\mathbf{i} + 16\mathbf{j}$

Magnitude $= \sqrt{7^2 + 16^2}$

$= 17.5$ N (to 3 sf);

(b) Angle $= \tan^{-1}\left(\dfrac{16}{7}\right)$

$= 66.4°$

EXERCISE 4C

1 Find the magnitude of the resultant of the following sets of forces, by forming a quadrilateral of forces.

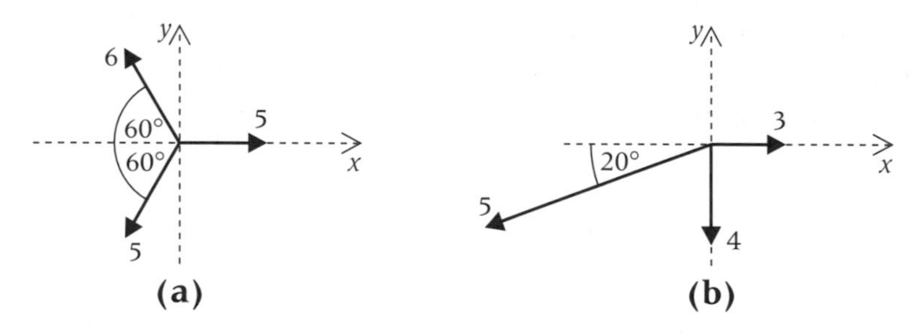

2 Find the magnitude of the resultant of the forces $(2\mathbf{i} + \mathbf{j} + \mathbf{k})$ N, $(3\mathbf{i} - 2\mathbf{j} + 3\mathbf{k})$ N and $(-2\mathbf{i} + \mathbf{k})$ N.

3 The resultant of the forces $(2\mathbf{i} + \mathbf{j})$ N, $3\mathbf{j}$ N, $(2\mathbf{i} + 4\mathbf{j})$ N, $(6\mathbf{i} + b\mathbf{j})$ N and $(a\mathbf{i} + \mathbf{j})$ N is $(3\mathbf{i} + 4\mathbf{j})$ N, where **i** and **j** are perpendicular unit vectors. Find a and b.

4 Three forces $(3\mathbf{i} + 5\mathbf{j})$ N, $(4\mathbf{i} + 11\mathbf{j})$ N, $(2\mathbf{i} + \mathbf{j})$ N act at a point. Given that **i** and **j** are perpendicular unit vectors find:

(a) the resultant of the forces in the form $a\mathbf{i} + b\mathbf{j}$,

(b) the magnitude of this resultant,

(c) the angle that the resultant makes with the unit vector **i**. [A]

5 Two forces $(3\mathbf{i} + 2\mathbf{j})$ N and $(-5\mathbf{i} + \mathbf{j})$ N act at a point. Find the magnitude of the resultant of these forces and determine the angle which the resultant makes with the unit vector **i**. [A]

6 Three forces $(\mathbf{i} + \mathbf{j})$ N, $(-5\mathbf{i} + 3\mathbf{j})$ N and $\lambda\mathbf{i}$ N, where $\mathbf{i}$ and $\mathbf{j}$ are perpendicular unit vectors, act at a point. Express the resultant in the form $(a\mathbf{i} + b\mathbf{j})$ and find its magnitude in terms of λ. Given that the resultant has magnitude 5 N, find the two possible values of λ.

Take the larger value of λ and find the tangent of the angle between the resultant and the unit vector $\mathbf{i}$. [A]

4.5 Resolving forces

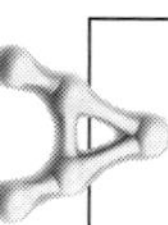

A force can be divided into two mutually perpendicular components whose vector sum is equal to the given force.

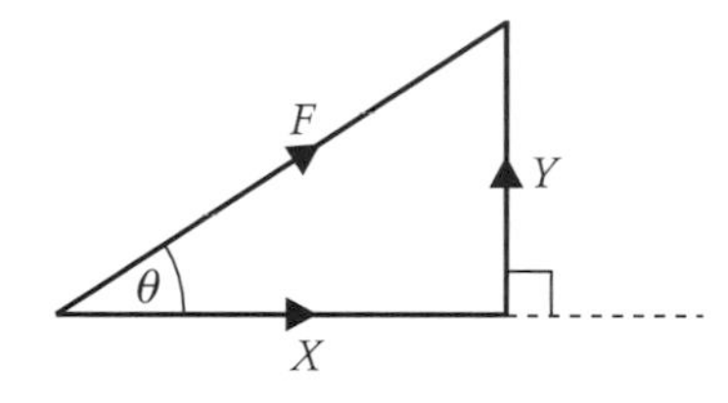

From the right-angled triangle:

$$X = F\cos\theta$$

and

$$Y = F\sin\theta$$

This process enables us to write forces in the form $a\mathbf{i} + b\mathbf{j}$. In the case above we would write:

$$F\cos\theta\,\mathbf{i} + F\sin\theta\,\mathbf{j}$$

Worked example 4.6

A force of 10 N acts at 60° below the horizontal. The unit vectors $\mathbf{i}$ and $\mathbf{j}$ are horizontal and vertical, respectively. Write the force in terms of the unit vectors $\mathbf{i}$, and $\mathbf{j}$.

Solution

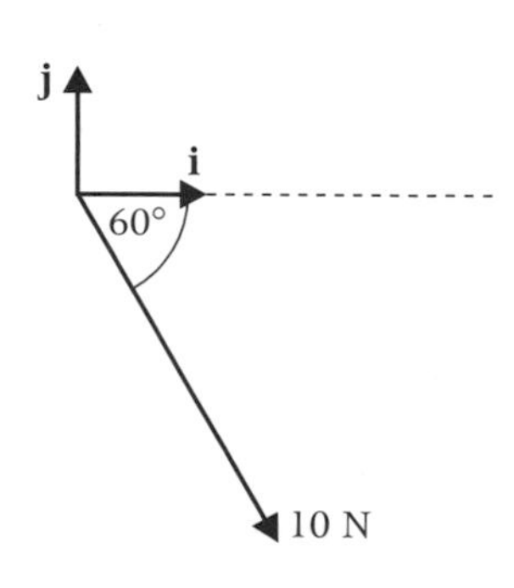

The diagram shows the force and the unit vectors.

The horizontal component of the force is 10 cos 60°.

The vertical component of the force is −10 sin 60°

Hence the force can be written as:

$$10\cos 60°\,\mathbf{i} - 10\sin 60°\,\mathbf{j} = 5\mathbf{i} - 8.66\mathbf{j} \text{ (to 3 sf)}$$

Worked example 4.7

Find the magnitude and direction of the resultant of the set of forces given below.

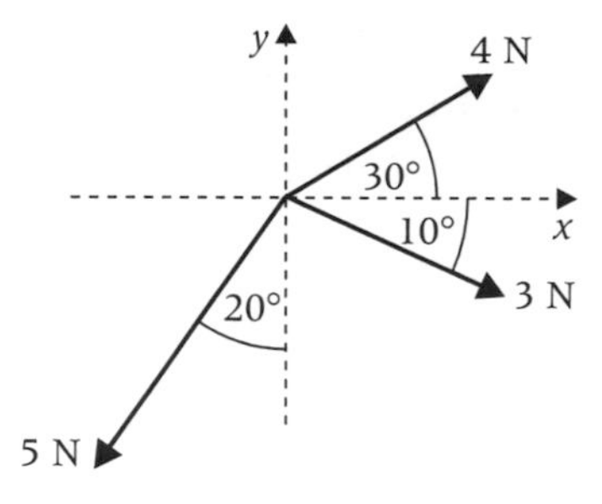

Solution

Resolving horizontally:

$$4\cos 30° + 3\cos 10° - 5\sin 20° = 4.708$$

Resolving vertically:

$$4\sin 30° - 3\sin 10° - 5\cos 20° = -3.219$$

The resultant force has magnitude $R = \sqrt{4.708^2 + 3.219^2} = 5.70$ N (to three significant figures) and acts at an angle

$\tan^{-1}\left(\dfrac{3.219}{4.708}\right) = 34.4°$ below the positive x-axis.

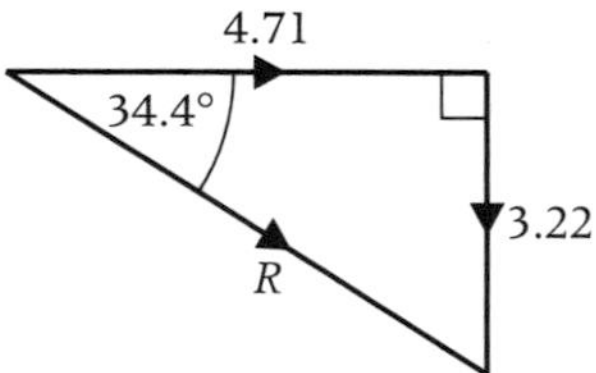

Examiner's tip: It is a good idea to draw a diagram in such cases, as it makes your intention clear

EXERCISE 4D

1 Find the components of the following forces in the directions of the x-axis and y-axis.

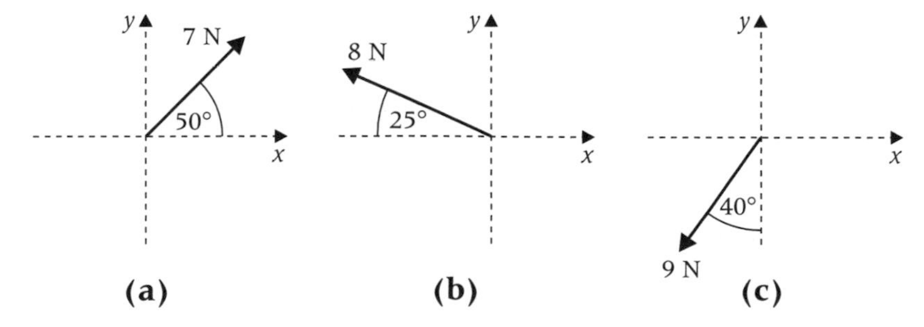

2 Express the forces in question **1** in the form $a\mathbf{i} + b\mathbf{j}$, if $\mathbf{i}$ is directed along the x-axis and $\mathbf{j}$ is directed along the y-axis.

3 Find the components of the following forces in the directions of the x-axis and y-axis.

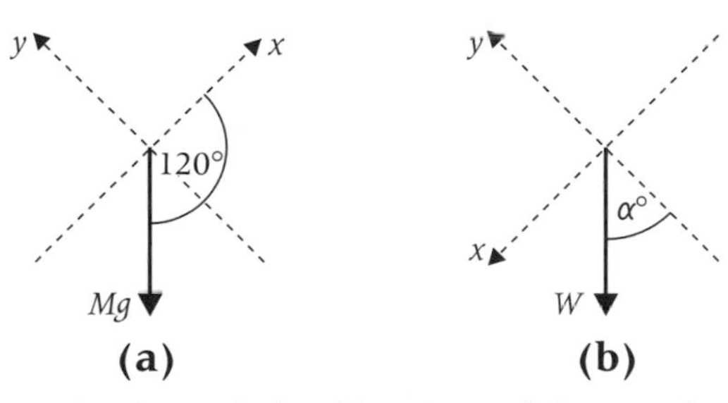

4 Find the magnitude and the direction of the resultant force in each of the following cases:

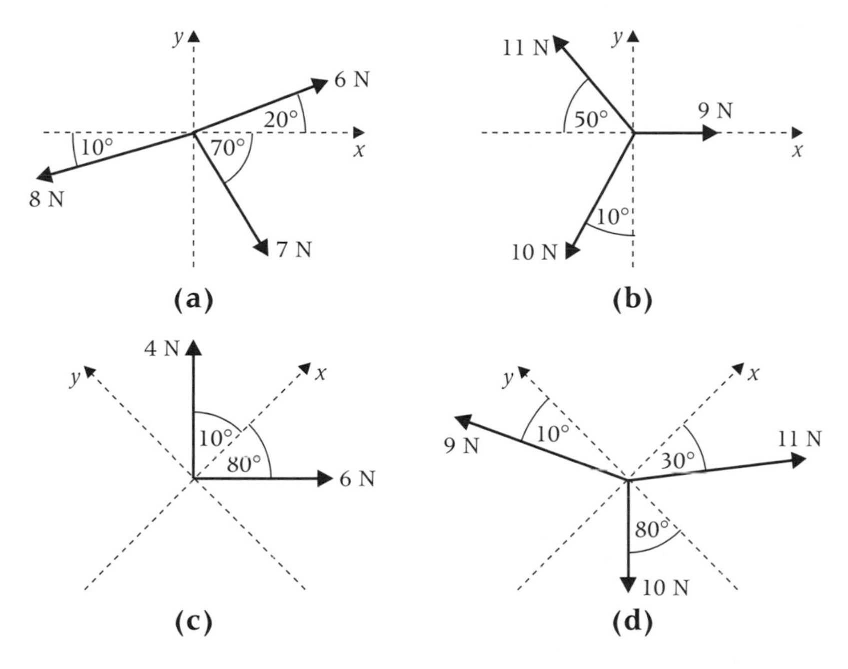

5 Three horizontal forces each of magnitude 10 N act in the directions of the bearings 040°, 160° and 280°. Find the magnitude of their resultant.

6 The following diagrams show a particle on an inclined plane. Find the components of the weight of the particle in the directions of the x-axis and y-axis.

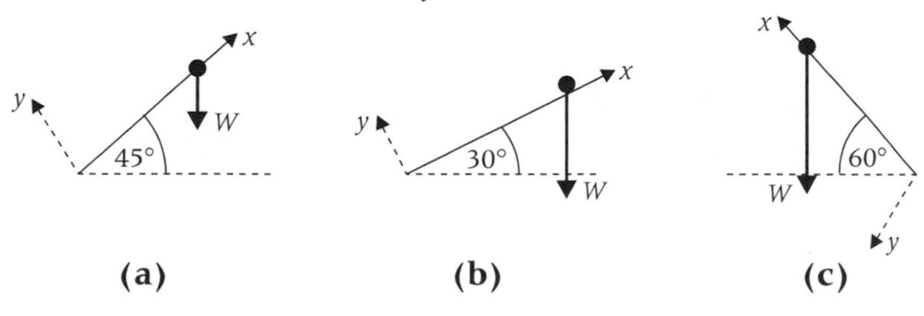

7 A particle of mass m kg lies at rest on a rough plane, inclined at α to the horizontal. What is the component of the weight down the plane?

4.6 Equilibrium

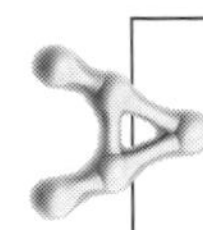

If a set of forces act on a particle such that their resultant is zero, then the particle is said to be in equilibrium: that is, no unbalanced forces act on the particle.

When a particle is in equilibrium it will either remain at rest or move with a constant velocity.

Two forces

If only two forces act on a particle, which is in equilibrium, then the forces must be equal and opposite. For example, a particle of mass m kg rests in equilibrium, suspended by a light inextensible string.

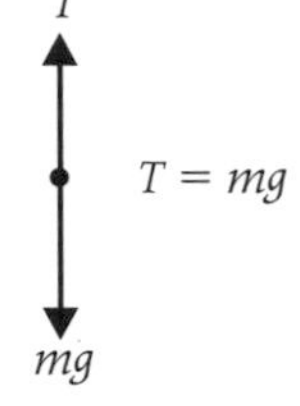

Three forces

If forces, of magnitude F_1, F_2 and F_2, have a zero resultant, then when the forces are placed end to end they must make a triangle.

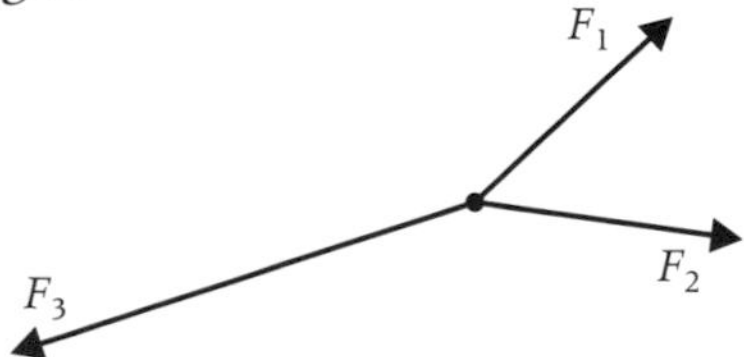

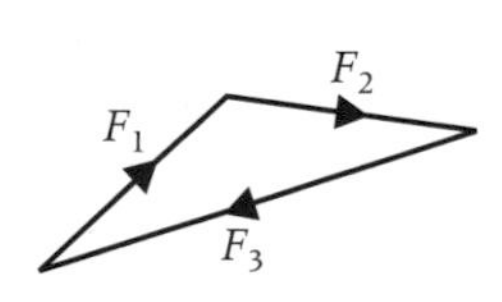

The sine rule and the cosine rule can be used in this triangle.

Worked example 4.8

A particle of mass 10 kg hangs in equilibrium suspended by two light inextensible strings, as shown. Find the tensions in the strings.

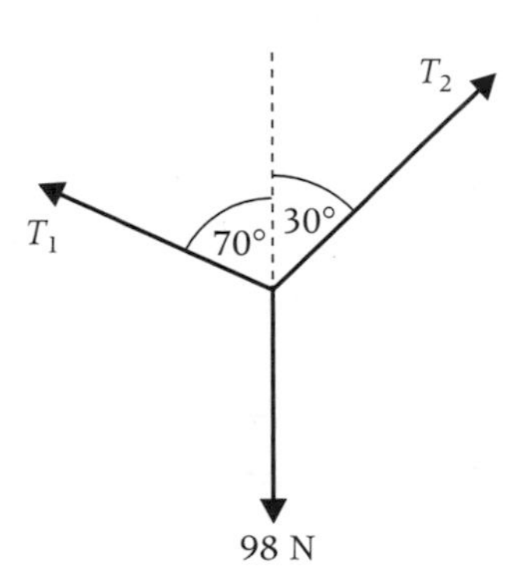

Solution

The diagram below shows how the forces must form a triangle, as they are in equilibrium.

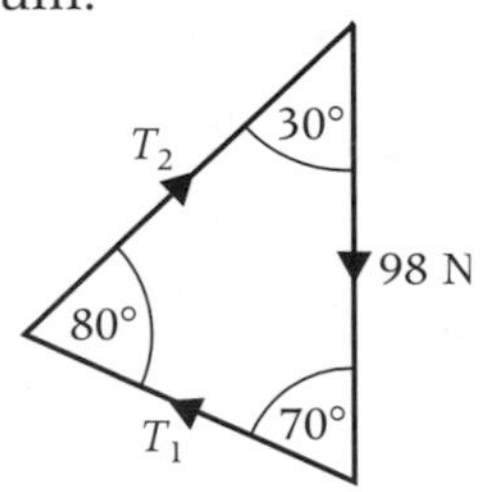

From the sine rule:

$$\frac{T_1}{\sin 30°} = \frac{T_2}{\sin 70°} = \frac{98}{\sin 80°}$$

$$T_1 = \frac{98 \sin 30°}{\sin 80°} \quad \text{and} \quad T_2 = \frac{98 \sin 70°}{\sin 80°}$$

$T_1 = 49.8$ N, $T_2 = 93.5$ N (to 3 sf)

Worked example 4.9

The set of forces shown below is in equilibrium. Find P and θ.

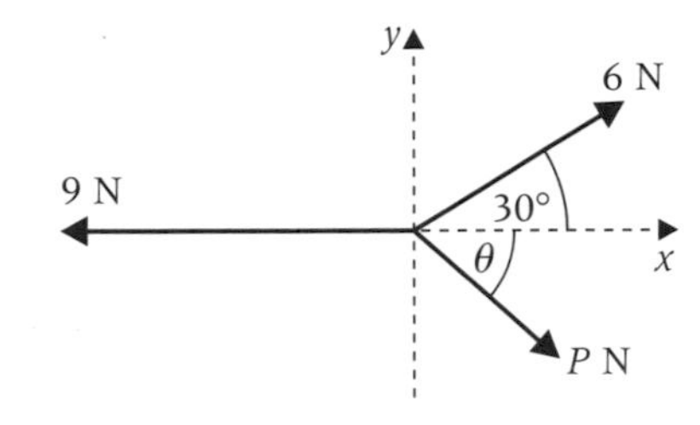

Solution

As the forces are in equilibrium, they can be arranged to form a triangle as shown in the diagram.

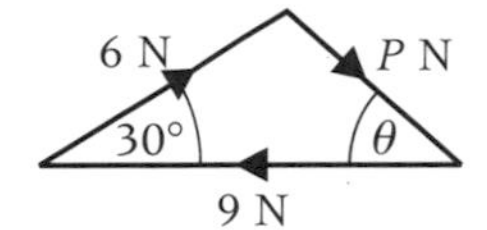

Using the cosine rule in this triangle gives:

$$P^2 = 6^2 + 9^2 - 2 \times 6 \times 9 \times \cos 30°$$

$$P = 4.84 \text{ N}$$

From the sine rule:

$$\frac{\sin \theta}{6} = \frac{\sin 30°}{4.84}$$

$$\theta = 38.3°$$

More than three forces

The previous two problems were solved using the sine and cosine rules only. Alternative solutions can be produced by resolving the forces in two perpendicular directions, for example horizontal and vertical. This method is particularly useful when we have more than three forces acting on the particle.

Worked example 4.10

The following set of forces is in equilibrium. Find P and the angle θ.

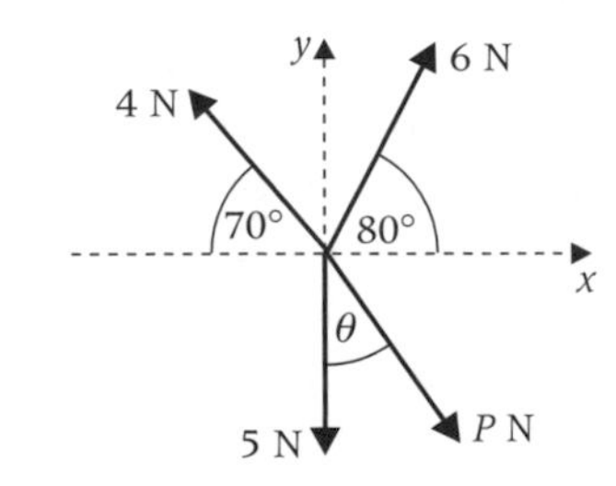

Solution

Resolving vertically:

$$P\cos\theta + 5 = 4\sin 70^\circ + 6\sin 80^\circ \qquad (1)$$

$$P\cos\theta = 4.667$$

Resolving horizontally:

$$P\sin\theta + 6\cos 80^\circ = 4\cos 70^\circ \qquad (2)$$

$$P\sin\theta = 0.326$$

Simultaneous equations like (1) and (2) occur frequently in mechanics. They can be solved using the trigonometric identities

'$\sin^2\theta + \cos^2\theta \equiv 1$' 'tan $\theta \equiv \dfrac{\sin\theta}{\cos\theta}$'

Hence $\tan\theta = \dfrac{0.326}{4.667}$

$$\theta = 4.0^\circ$$

and $P^2 = 4.667^2 + 0.326^2$

$$P = 4.68\text{ N}$$

Worked example 4.11

A particle of mass 5 kg is at rest on a rough plane inclined at 50° to the horizontal. Find the normal reaction and the frictional force on the particle.

Solution

The diagram shows the forces acting on the particle.

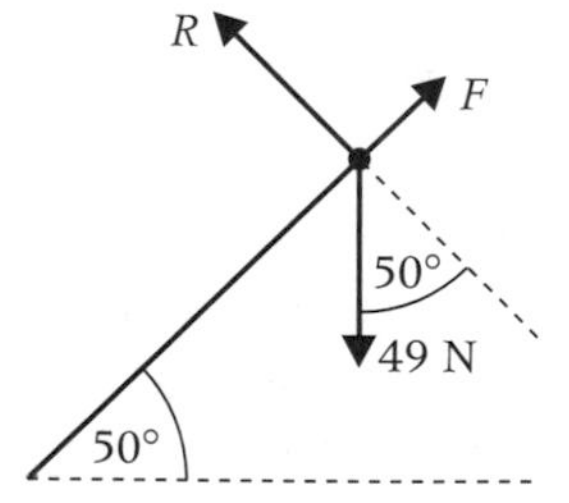

Resolving parallel to the plane:

$F = 49\sin 50^\circ = 37.5$ N

Resolving perpendicular to the plane:

$R = 49\cos 50^\circ = 31.5$ N

Examiner's tip: The choice of whether to solve a problem by resolving or by drawing the triangle of forces and using trigonometry depends upon the particular situation involved. In the preceding examples, two demonstrated the use of trigonometry, and two showed how to find the solution by resolving. Whereas some problems are much easier done one way than the other, either technique will work equally well for certain types of problem. Worked example 4.11, for instance could be solved equally well by either method.

EXERCISE 4E

1 Each of the following sets of forces is in equilibrium. Find the magnitudes F_1 and F_2.

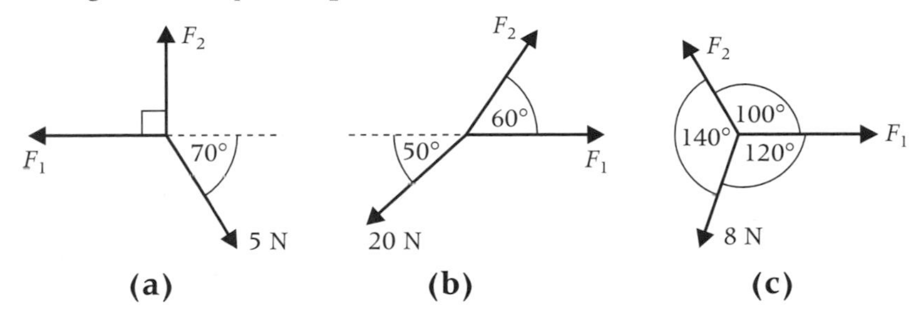

2 Each of the following sets of forces is in equilibrium. Find F and the angle θ.

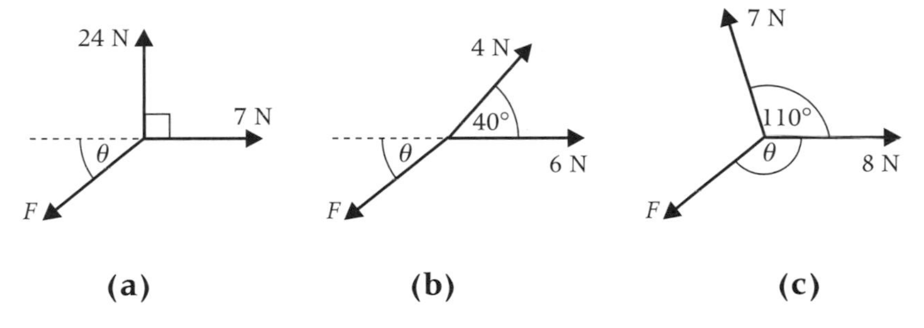

3 Each of the following sets of forces is in equilibrium. Find F_1 and F_2.

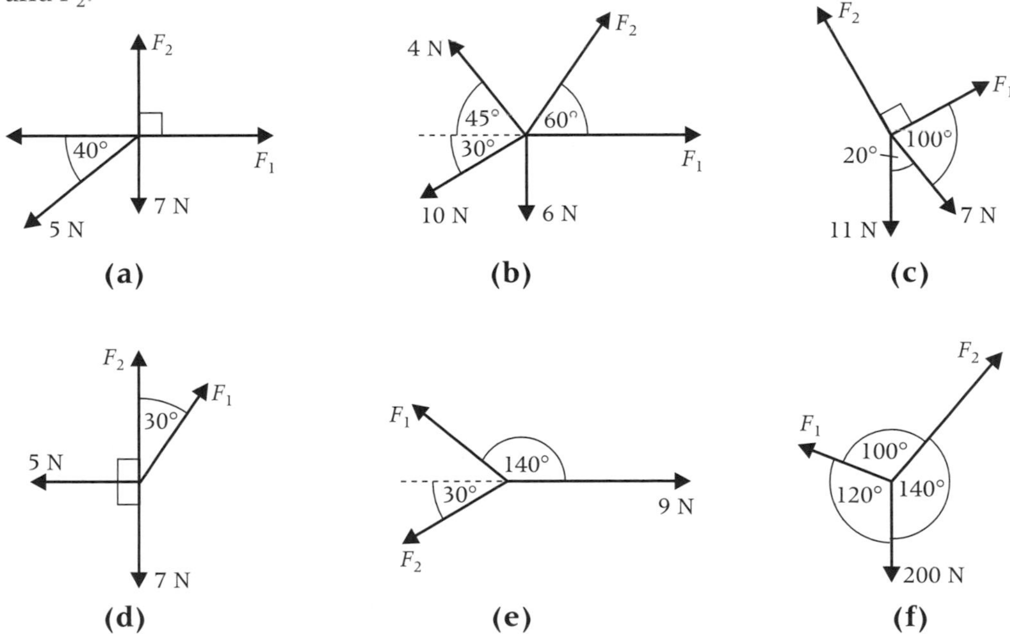

4 A particle is in equilibrium, subject to forces $(6\mathbf{i} + \mathbf{j})$ N, $(2\mathbf{i} + 3\mathbf{j})$ N and $\mathbf{P}$.

(a) Find $\mathbf{P}$ in terms of $\mathbf{i}$ and $\mathbf{j}$.

(b) Find the magnitude of $\mathbf{P}$ and the angle between $\mathbf{P}$ and $\mathbf{i}$.

5 The set of forces $(a\mathbf{i} + 2\mathbf{j})$ N, $(6\mathbf{i} - 3\mathbf{j})$ N, $(4\mathbf{i} + b\mathbf{j})$ N, and $(\mathbf{i} + \mathbf{j})$ N are in equilibrium. Find a and b.

6 Forces $\mathbf{F}_1$, $\mathbf{F}_2$ and $\mathbf{F}_3$ are in equilibrium.

(a) If $\mathbf{F}_1 = (\mathbf{i} + \mathbf{j})$ N and $\mathbf{F}_3 = (7\mathbf{i} + 8\mathbf{j})$ N, find $\mathbf{F}_2$.

(b) If $\mathbf{F}_1 = (2\mathbf{i} + \mathbf{j})$ N and $\mathbf{F}_2 = (-4\mathbf{i} + \mathbf{j})$ N, find $\mathbf{F}_3$.

7 Horizontal forces each of magnitude 10 N act in the direction of the bearings 040°, 160° and 280°. Are these forces in equilibrium?

8 A mass of 5 kg is suspended by two light, inextensible strings. The angles between the strings and the vertical are 30° and 60°. Find the tensions in the strings.

9 A mass of 10 kg is suspended by two light, inextensible strings. The angles between the strings and the vertical are 40° and 20°. Find the tensions in the strings.

10 A particle of weight 10 N is suspended by two light, inextensible strings. The tension in one string is 5 N, which acts at 20° to the vertical. Find the tension in the second string and the angle between it and the vertical.

11 A particle of weight 10 N is in equilibrium on a smooth plane, inclined 40° to the horizontal. A horizontal force P acts on the particle. Find P and the normal reaction between the plane and the particle.

12 A particle of weight 10 N is suspended from a fixed point by a light inextensible string. A horizontal force of 5 N also acts on the particle. Find the tension in the string and the angle between the string and the vertical.

13 Three forces act upon a particle, which is in equilibrium. If the magnitudes of the forces are 4 N, 5 N, and 6 N, find the angles between the forces.

14 A and B are particles connected by a light, inextensible string which passes over a smooth fixed pulley attached to a corner of a smooth fixed plane, inclined at 37° to the horizontal. Particle B hangs freely. If the mass of A is 3 kg, and the system is in equilibrium, find the mass of B.

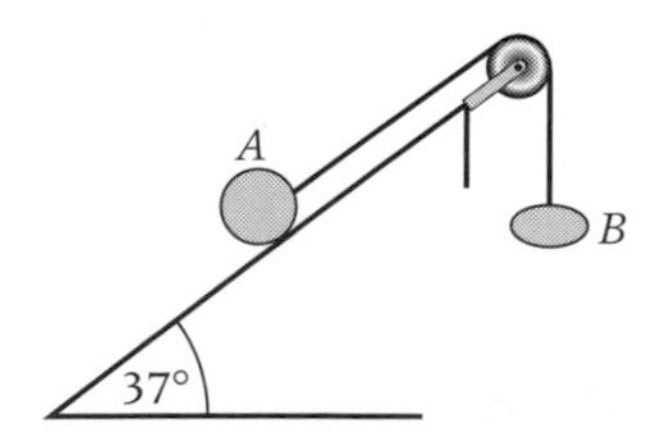

15 An object of mass 3 kg is suspended by two light, inextensible strings. The strings make angles of 30° and 40° and the horizontal, as shown in the diagram.

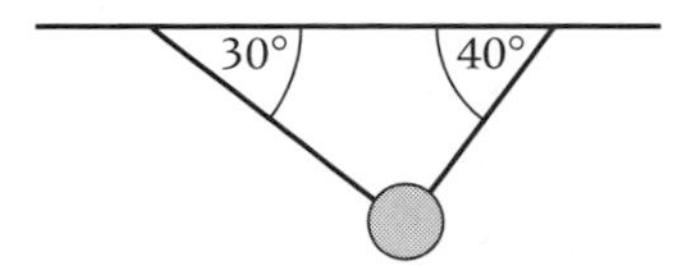

Find the magnitude of the tension in each string.

16 An object, of mass 40 kg, is supported in equilibrium by four cables. The forces, in newtons, exerted by three of the cables, $\mathbf{F}_1$, $\mathbf{F}_2$ and $\mathbf{F}_3$ are given in terms of the unit vectors **i**, **j** and **k** as $\mathbf{F}_1 = 80\mathbf{i} + 20\mathbf{j} + 100\mathbf{k}$, $\mathbf{F}_2 = 60\mathbf{i} - 40\mathbf{j} + 80\mathbf{k}$ and $\mathbf{F}_3 = -50\mathbf{i} - 100\mathbf{j} + 80\mathbf{k}$. The unit vectors **i** and **j** are perpendicular and horizontal and the unit vector **k** is vertically upwards.

Find $\mathbf{F}_4$, the force exerted by the fourth cable, in terms of **i**, **j** and **k**. Also find its magnitude to the nearest newton. [A]

17 Four boys are playing a 'tug of war' game, each pulling horizontally on a rope attached to a light ring. Boy *A* pulls with a force of $(92\mathbf{i} - 33\mathbf{j})$ N, boy *B* with force $(66\mathbf{i} + 62\mathbf{j})$ N and by *C* with force $(-70\mathbf{i} + 99\mathbf{j})$ N, where **i** and **j** are perpendicular unit vectors. Given that the ring is in equilibrium, find the force exerted by boy *D*, and its magnitude.

4.7 Friction

Consider a body of mass m, placed on a rough horizontal plane. Suppose a horizontal force of magnitude P is applied. If the body remains in equilibrium, then the frictional force, F, will be such that $F = P$.

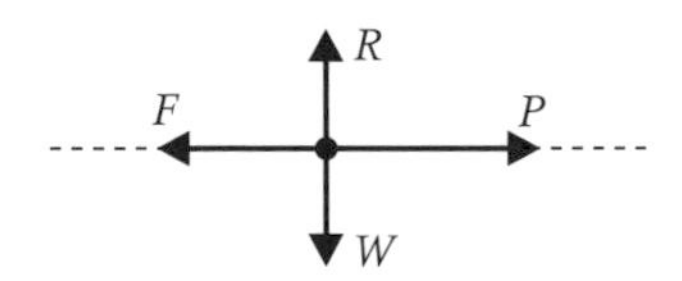

If P is steadily increased and the body remains at rest, then F increases also, so that $F = P$ remains true. However F can only increase up to a certain limit, F_{max}.

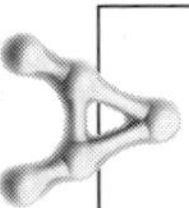

> The force between the surfaces in contact and the type of surface determine how large the frictional force can become.

Racing cars make use of this principle. The downwards force on the vehicle is increased by incorporating an aerofoil into the design. This improves the grip of the tyres on the road surface, so that drivers can take corners at greater speed without skidding.

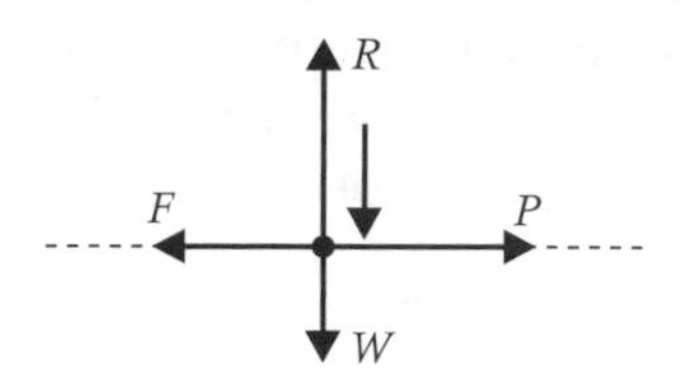

The vertical force between the surfaces is the normal reaction R. If a weight is placed on top of a body so that the normal reaction is doubled, it will be found that F_{max} will also double. That is, F_{max} is proportional to R (the normal reaction).

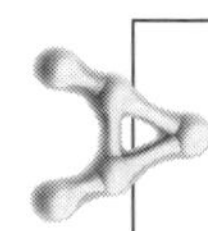

$$F_{max} = \mu R$$

where μ, a constant, which depends only on the roughness of the surface, is known as the coefficient of friction.

If P is increased even further then slipping will occur. The frictional force cannot increase further and experiments show that the frictional force remains constant at its maximum throughout the motion.

Note that: **(i)** When the frictional force does equal its maximum it is often said to be **limiting**; **(ii)** The special case $\mu = 0$ means $F_{max} = 0$, which corresponds to a smooth plane.

Horizontal planes

Worked example 4.12

A particle of mass 10 kg is placed on a rough horizontal plane. A horizontal force P acts on the particle. P is increased until the particle is on the point of sliding, which occurs when $P = 10$ N. Find the coefficient of friction between the particle and the plane.

Solution

The diagram shows the forces acting on the particle.

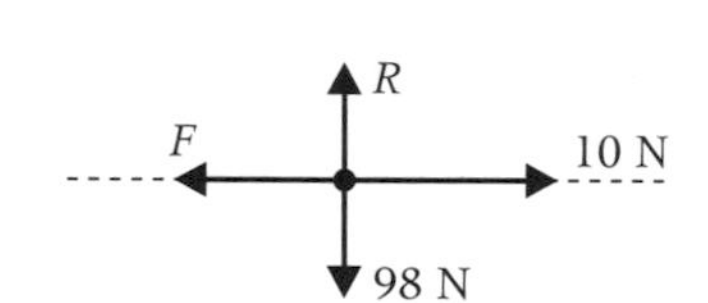

Resolving vertically: $R = 98$ N

Resolving horizontally: $F = 10$ N

If motion is just about to occur, friction is limiting, $F = F_{max}$ so $F = \mu R$.

$$10 = 98\mu$$

$$\mu = \frac{5}{49}.$$

Worked example 4.13

In the following situations a body of mass 10 kg is placed on a rough horizontal plane. If $\mu = \frac{1}{2}$ in each case determine whether motion will occur.

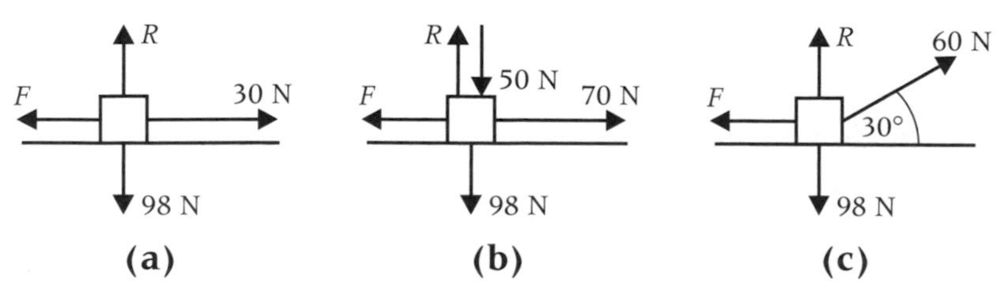

Solution

(a) Resolving vertically: $R = 98$ N

$F_{max} = 0.5 \times 98 = 49$ N

For equilibrium $F = 30 < 49$, so no motion will occur.

(b) Resolving vertically: $R = 98 + 50 = 148$ N

$F_{max} = 0.5 \times 148 = 74$ N

For equilibrium $F = 70 < 74$, so again no motion will occur.

(c) Resolving vertically: $R + 60 \sin 30° = 98$

$R = 68$ N

$F_{max} = 0.5 \times 68 = 34$ N

For equilibrium $F = 60 \cos 30° = 52 > 34$, which is not possible, so sliding will occur.

Use of the inequality $F \leqslant \mu R$

In all possible cases of a particle in equilibrium on a rough plane, the frictional force has a limiting value so the following inequality is always true:

$$F \leqslant \mu R$$

This inequality itself can be used to good effect.

Worked example 4.14

A particle, of mass 10 kg, is at rest on a rough horizontal plane. A force, of magnitude of P N, is applied in the direction shown. If $\mu = \frac{1}{2}$, what is the greatest possible value of the magnitude of P such that motion does not occur?

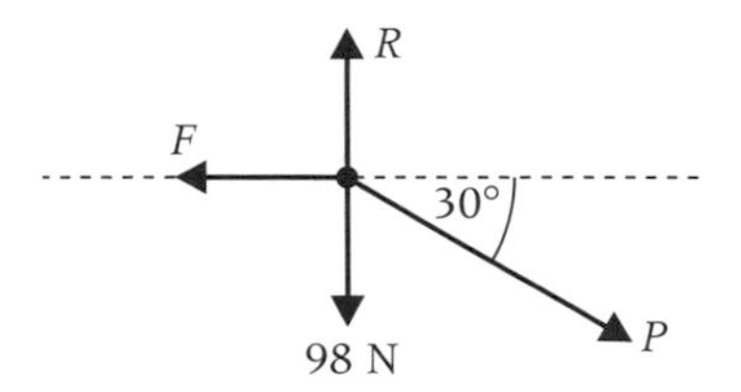

Solution

Resolving horizontally: $F = P \cos 30°$.

Resolving vertically: $R = 98 + P \sin 30°$.

If we substitute for F and R in '$F \leqslant \mu R$'

$$P \cos 30° \leqslant \tfrac{1}{2}(98 + P \sin 30°)$$

Rearranging gives:

$$P(\cos 30° - \tfrac{1}{2} \sin 30°) \leqslant 49$$

$$P \leqslant \frac{49}{\cos 30° - \frac{1}{2} \sin 30°}$$

$$P \leqslant 79.5 \text{ N}$$

So the greatest value of P is 79.5 N (to 3 sf)

Worked example 4.15

A particle of weight W is at rest on a rough horizontal surface. A force of magnitude W is applied to the particle as shown. Show that the least value of μ is $1 + \sqrt{2}$.

Solution

The diagram shows the forces acting on the diagram.

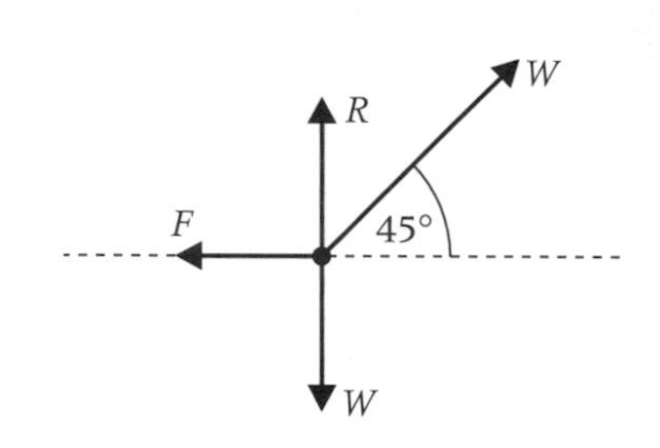

Resolving horizontally: $F = W\cos 45° = \dfrac{\sqrt{2}}{2}W$

Resolving vertically: $R + W\sin 45° = W$

$$R = W - W\sin 45°$$
$$= W\left(1 - \frac{\sqrt{2}}{2}\right)$$

Substituting for F and R in $F \leqslant \mu R$ we get:

$$\frac{\sqrt{2}}{2}\text{W} \leqslant \mu\text{W}(1 - \frac{\sqrt{2}}{2})$$

$$\frac{\sqrt{2}}{2\left(1 - \frac{\sqrt{2}}{2}\right)} \leqslant \mu$$

Rationalising the denominator gives:

$$\frac{\sqrt{2}(2 + \sqrt{2})}{(2 - \sqrt{2})(2 + \sqrt{2})} \leqslant \mu$$

$$\frac{2\sqrt{2} + 2}{2} \leqslant \mu$$

$$1 + \sqrt{2} \leqslant \mu$$

So the least value of μ which satisfies this inequality is $1 + \sqrt{2}$.

EXERCISE 4F

1 A horizontal force of magnitude 5 N is applied to a body of mass 20 kg which is at rest on a rough horizontal plane. Find the coefficient of friction, given that friction is limiting in this position.

2 In the following situations a particle of mass 4 kg is placed on a rough horizontal plane. If $\mu = 5/7$ determine whether motion will occur.

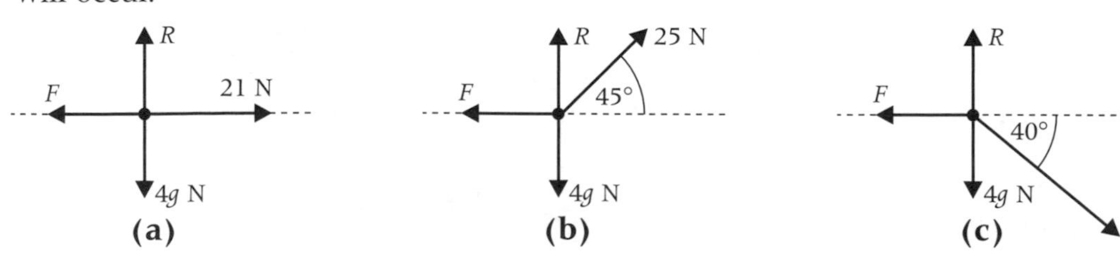

3 A particle of weight 10 N is at rest on a rough horizontal plane. The particle is pulled by a light, inextensible string, inclined at an angle of 20° to the plane. If the tension in the string is 5 N, find the least value of the coefficient of friction correct to 3 significant figures.

4 The diagram shows a particle of mass of 6 kg at rest on a rough horizontal plane, subject to an external force, of magnitude P N. What is the greatest value of P, if $\mu = 2/3$.

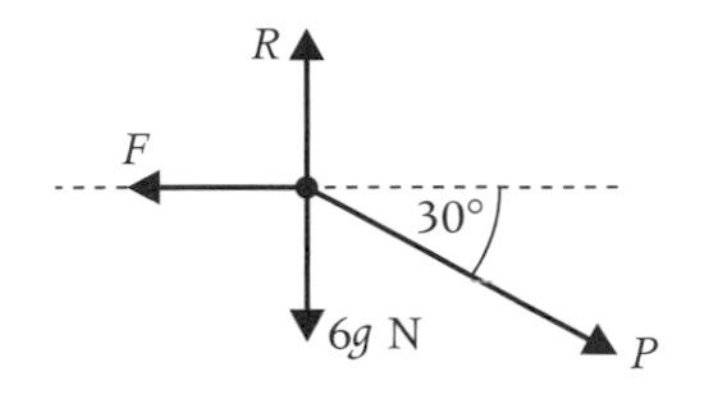

5 Horizontal forces, each of magnitude 100 N act in the direction of the bearings 050°, 170°, and 260°. Find the resultant of these forces. If these forces act on a particle, of mass 10 kg, which is in equilibrium on a rough horizontal plane, find the magnitude of frictional force, which acts on the particle. Find also the least value of the coefficient of friction.

6 Two particles A and B are attached by a light, inextensible string passing over a smooth fixed pulley, as shown, and A has twice the mass of B. Find the smallest value of μ, the coefficient of friction, for equilibrium to be maintained.

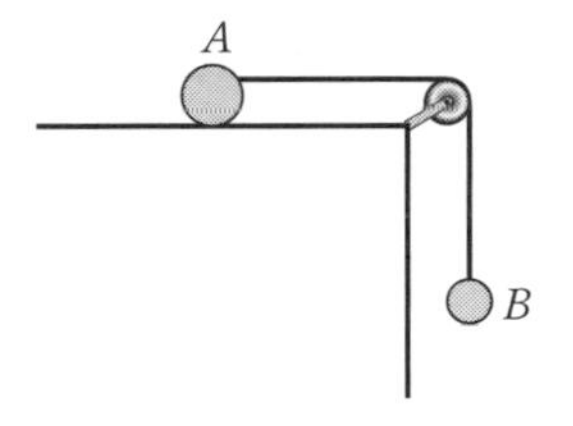

7 The coefficient of friction between a sledge and a snowy surface is 0.2. The combined mass of a child and the sledge is 45 kg. What is the least horizontal force necessary to pull the sledge along the horizontal surface at a constant speed?

8 A sledge of mass 12 kg is on level ground.

(a) A horizontal force of 10 N will keep the sledge moving at a constant speed. Find the value of the coefficient of friction.

(b) A girl of mass 25 kg sits on the sledge. Find the least horizontal force required to keep the sledge moving at a constant speed.

9 A small ring, of weight w N, is threaded on a horizontal curtain rail. A light, inextensible string is pulling it along the rail. The tension in the string is equal to $2w$ N. Show that the least value of the coefficient of friction is $2 - \sqrt{2}$.

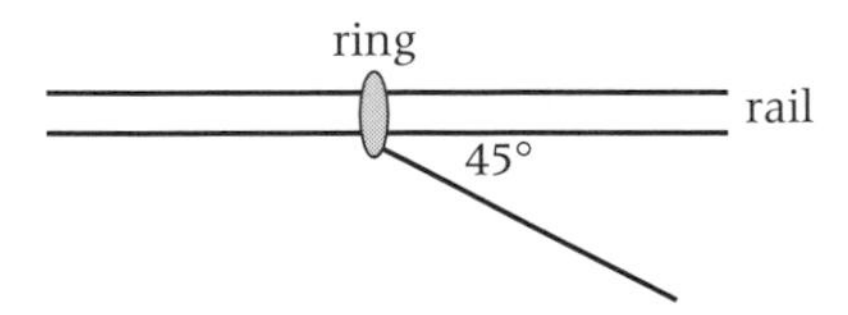

10 The coefficient of friction between a particle, of weight w, and a horizontal plane is μ. The particle is in equilibrium subject to a force of magnitude P N which acts at an angle of θ below the horizontal. Show that:

$$P \leqslant \frac{\mu w}{\cos\theta - \mu\sin\theta}$$

4.8 Friction on inclined planes

Consider a particle, of mass m kg, which is at rest on a rough plane inclined at an angle α to the horizontal.

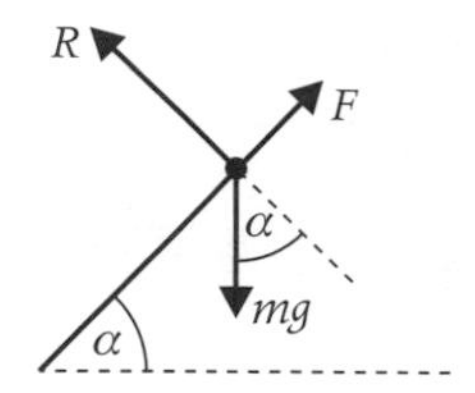

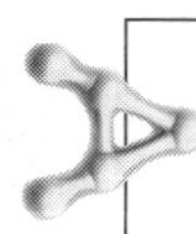

> The diagram shows the forces acting on the particle. The friction force must act up the slope to maintain equilibrium.

Resolving parallel to the plane: $F = mg \sin \alpha$

Resolving perpendicular to the plane: $R = mg \cos \alpha$

However

$$F \leqslant \mu R$$

so

$$mg \sin \alpha \leqslant \mu mg \cos \alpha$$

hence:

$$\frac{\sin \alpha}{\cos \alpha} \leqslant \mu$$

$$\tan \alpha \leqslant \mu$$

This result tells us that the particle will slip if the angle becomes too steep. The particle will be on the point of slipping when

$$\tan \alpha = \mu$$

or

$$\alpha = \tan^{-1} \mu$$

Worked example 4.16

A block of mass 3 kg is placed on a rough horizontal table. The table is gradually tilted until the particle begins to slip. The block is on the point of slipping when the table is inclined at an angle of 41° to the horizontal. Find the coefficient of friction.

Solution

The diagram shows the forces acting on the block.

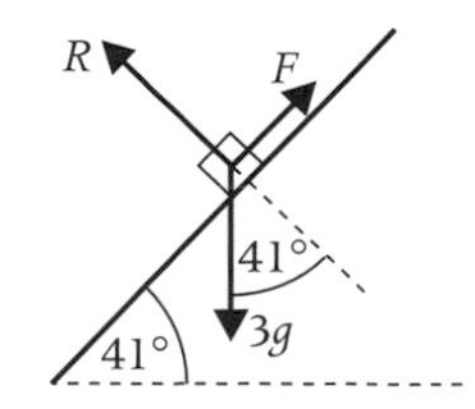

Resolving parallel to the plane: $F = 3g \sin 41° = 19.3$ N

Resolving perpendicular to the plane $R = 3g \cos 41° = 22.2$ N

But if the particle is on the point of sliding then friction is limiting so

$$F = \mu R$$
$$19.3 = 22.2\mu$$
$$\mu = 0.869 \text{ (to 3 sf)}.$$

Note: We could have used $\mu = \tan^{-1}\alpha$

Suppose this table is now placed at an angle of 50° to the horizontal. The block would not be able to rest in equilibrium unaided. So, suppose a force, of magnitude P, is applied to the block so that it acts up the plane. We will now find the least value of P to maintain equilibrium.

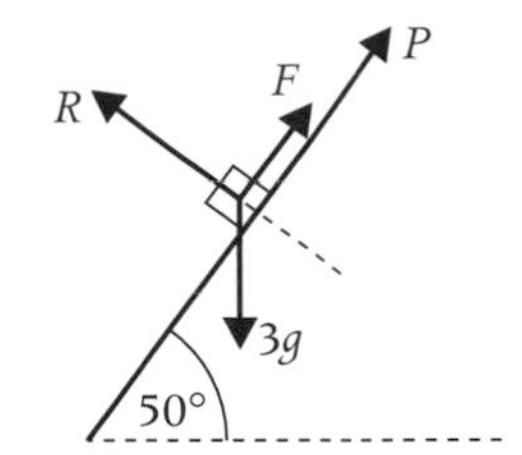

Resolving parallel to plane: $F + P = 3g \sin 50°$

$F = 3g \sin 50° - P$

Resolving perpendicular to plane: $R = 3g \cos 50°$

Here we can use $F \leqslant \mu R$, and substitute for F and R.

$$3g \sin 50° - P \leqslant \mu 3g \cos 50°$$
$$P \geqslant 3g \sin 50° - \mu 3g \cos 50°$$
$$P \geqslant 6.09 \text{ N (to 3 sf)}.$$

So the **minimum** value of P is 6.09 N if the particle is to remain at rest.

If the force P is too large, however, the particle will be pulled up the plane so we will now find the **greatest** value of P such that the particle will remain in equilibrium.

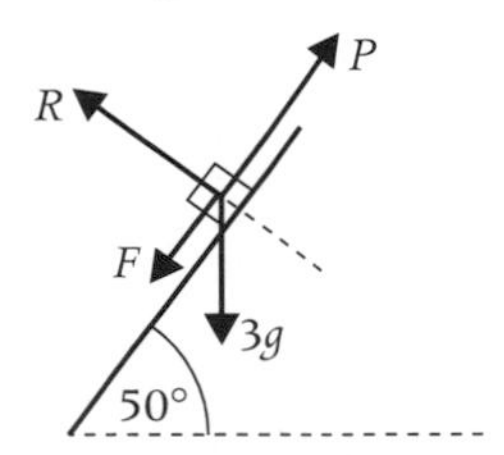

The method used to solve this problem will be exactly the same as that used to find the minimum value of P. There is one fundamental difference, however, which can be seen in the diagram; the frictional force now acts downhill. If we are considering pulling the particle uphill, then friction will oppose the motion and so act downhill.

4

Resolving parallel to plane: $P = F + 3g \sin 50°$

Resolving perpendicular to plane: $R = 3g \cos 50°$

Substituting into $F \leqslant \mu R$

$$P - 3g \sin 50° \leqslant \mu\, 3g \cos 50°$$

$$P \leqslant 3g \sin 50° + \mu\, 3g \cos 50°$$

$$P \leqslant 38.9 \text{ N}$$

Worked example 4.17

A body of weight W is placed on a rough plane, inclined at 30° to the horizontal, where $\mu = \frac{1}{\sqrt{3}}$. Find the greatest horizontal force that can be applied to the body, if it is to remain at rest.

Solution

The diagram shows the forces acting on the body.

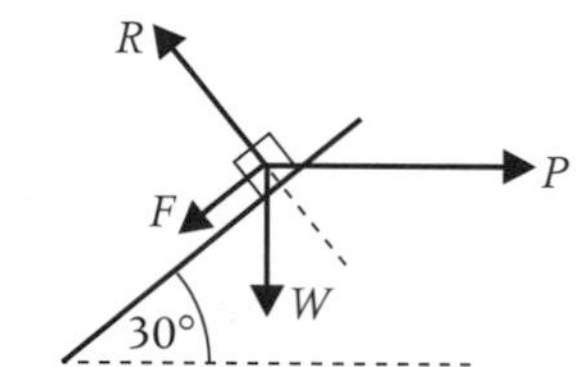

Resolving parallel to the plane: $P \cos 30° = F + W \sin 30°$

$$F = \frac{\sqrt{3}}{2}P - \frac{1}{2}W$$

Resolving perpendicular to the plane: $R = W \cos 30° + P \sin 30°$

$$R = \frac{\sqrt{3}}{2}W + \frac{1}{2}P$$

Substituting for F and R in $F \leqslant \mu R$ gives:

$$\frac{\sqrt{3}}{2}P - \frac{1}{2}W \leqslant \frac{1}{\sqrt{3}}\left(\frac{1}{2}P + \frac{\sqrt{3}}{2}W\right)$$

$$\frac{\sqrt{3}}{2}P - \frac{1}{2\sqrt{3}}P \leqslant W$$

$$\frac{\sqrt{3}}{2}P - \frac{\sqrt{3}}{6}P \leqslant W$$

$$\frac{\sqrt{3}}{3}P \leqslant W$$

$$P \leqslant \sqrt{3}W$$

So the greatest force which can be applied is $\sqrt{3}W$.

EXERCISE 4G

1 In the following situations a particle of mass m kg is placed on a rough plane inclined at an angle α to the horizontal. Determine whether the particle will slide.

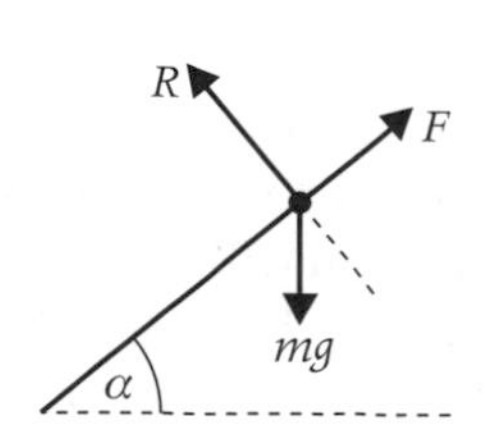

(a) $\alpha = 20°, \mu = 0.1,$

(b) $\alpha = 30°, \mu = 0.75,$

(c) $\alpha = 50°, \mu = 0.9.$

2 The situations below show a particle of mass 10 kg at rest on an inclined plane, where the coefficient of friction is 0.1. Find the magnitude, P, of the applied force, if the particle is on the point of sliding down the plane.

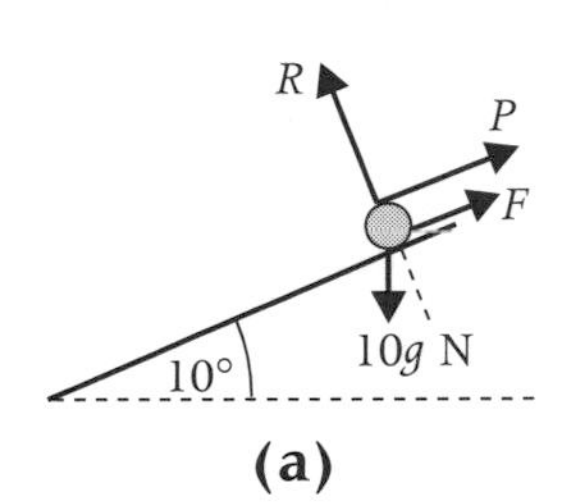

(a)

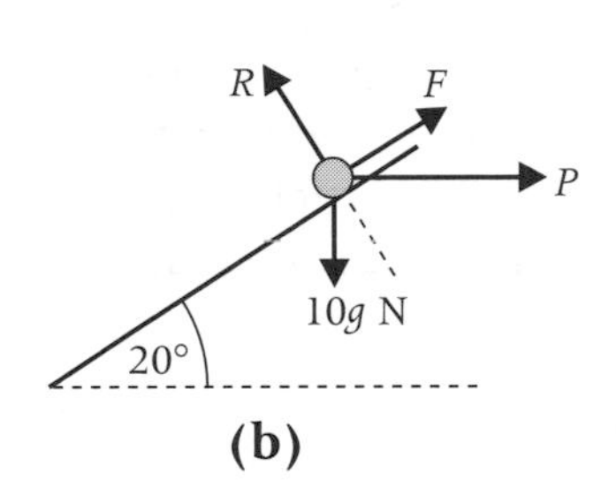

(b)

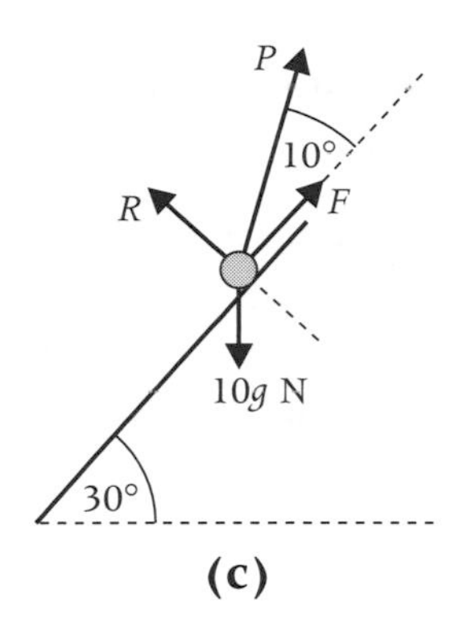

(c)

3 The situations below show a particle of mass 10 kg at rest on a rough inclined plane, where the coefficient of friction is 0.1. Find the magnitude, P, of the force applied to the particle, if it is on the point of moving up the plane.

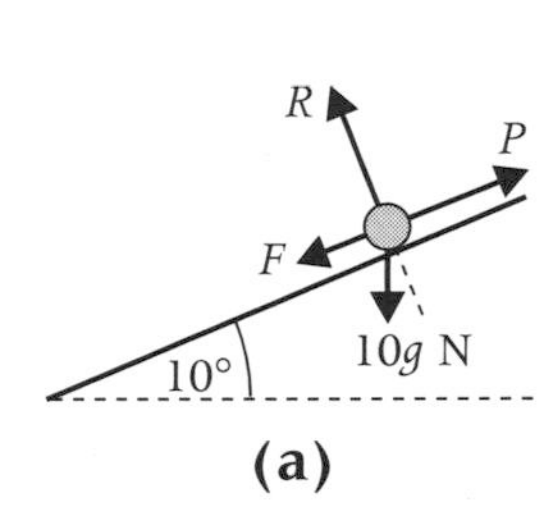

(a)

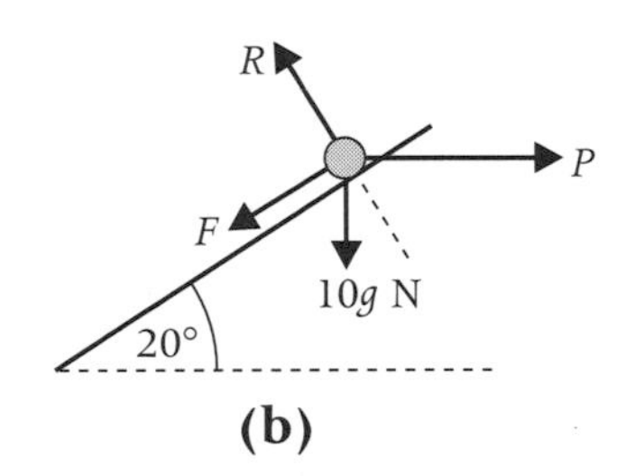

(b)

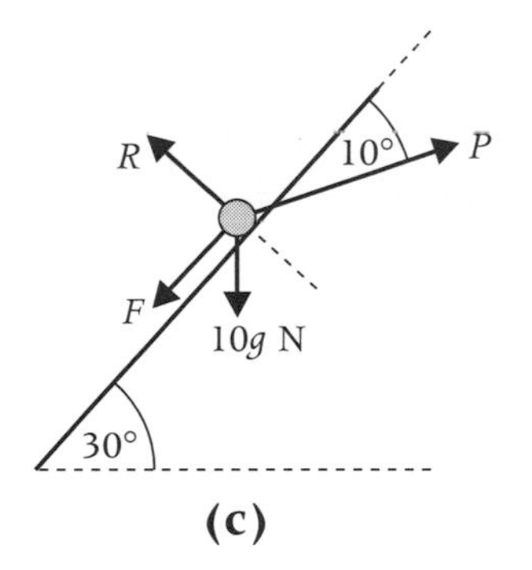

(c)

4 A particle of mass 10 kg is placed on a rough plane, inclined at 45° to the horizontal. If $\mu = 0.5$ find the least force required to keep the particle in equilibrium, if the force acts upwards, along the line of greatest slope.

5 A particle of mass 6 kg is at rest on a rough plane, of inclination α to the horizontal. Find the greatest horizontal force that can be applied to the particle, if it is to remain in equilibrium, in each of the following cases:

(a) $\alpha = 20°$, $\mu = 0.1$,

(b) $\alpha = 30°$, $\mu = 0.9$,

(c) $\alpha = 50°$, $\mu = 0.7$.

6 A particle, of mass 12 kg, is at rest on a rough plane, inclined at an angle α to the horizontal. Find the least force, which is parallel to the plane, that must be applied to the particle, if it is to remain in equilibrium, in each of the following cases:

(a) $\alpha = 20°$, $\mu = 0.1$,

(b) $\alpha = 30°$, $\mu = 0.5$,

(c) $\alpha = 50°$, $\mu = 1.0$.

7 A particle of mass 8 kg is at rest on a rough plane inclined at an angle of 30° to the horizontal. A horizontal force of 20 N acts on the particle as shown. Find the magnitude of the friction force and the normal reaction on the particle. What is the least value of μ?

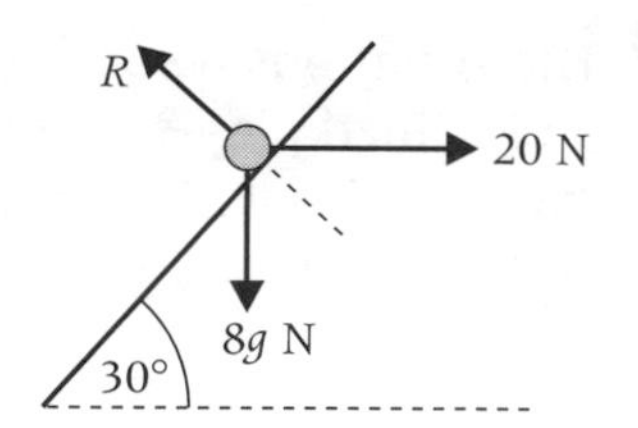

8 The situations below show a particle of mass 4 kg at rest on an inclined plane, subject to a given external force. Draw a force diagram showing all the forces acting on the particle. Find the normal reaction, the friction force, and the range of values of the coefficient of friction in each case.

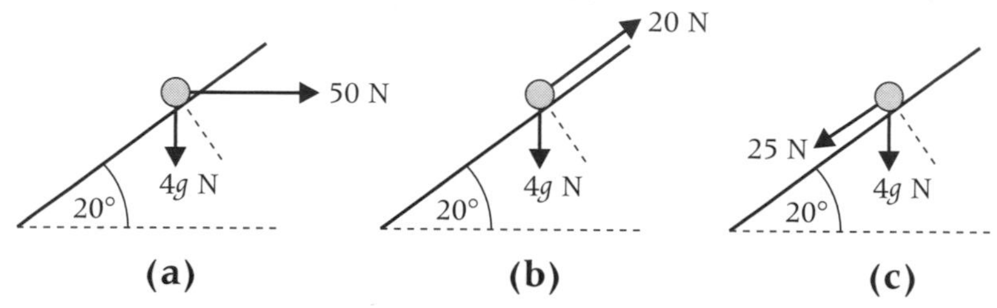

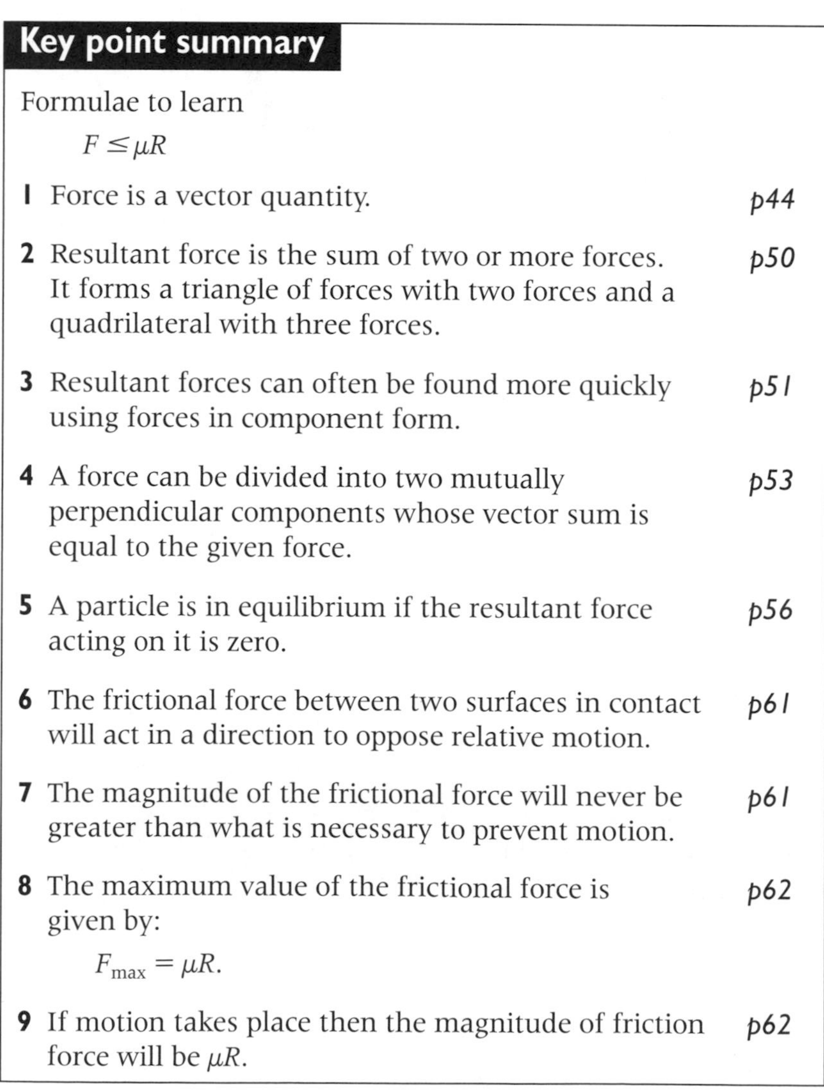

Key point summary

Formulae to learn

$F \leq \mu R$

1 Force is a vector quantity. *p44*

2 Resultant force is the sum of two or more forces. It forms a triangle of forces with two forces and a quadrilateral with three forces. *p50*

3 Resultant forces can often be found more quickly using forces in component form. *p51*

4 A force can be divided into two mutually perpendicular components whose vector sum is equal to the given force. *p53*

5 A particle is in equilibrium if the resultant force acting on it is zero. *p56*

6 The frictional force between two surfaces in contact will act in a direction to oppose relative motion. *p61*

7 The magnitude of the frictional force will never be greater than what is necessary to prevent motion. *p61*

8 The maximum value of the frictional force is given by: *p62*

$F_{max} = \mu R.$

9 If motion takes place then the magnitude of friction force will be μR. *p62*

Test yourself	What to review
1 Draw diagrams to show the forces acting on each of the following. You should include air resistance where appropriate, and model each body as a particle. **(a)** A golf ball at its maximum height. **(b)** A cyclist travelling up a slope at a constant speed. **(c)** A child on a swing at her lowest position.	*Section 4.2*
2 The diagram shows three forces and the perpendicular unit vectors **i** and **j**. **(a)** Find the resultant of these three forces in terms of the unit vectors **i** and **j**. **(b)** Find the magnitude of the resultant of these three forces and draw a diagram to show the direction in which it acts. **(c)** When a fourth force acts at the same point the forces are in equilibrium. Find the magnitude of this force and describe the direction in which it acts.	*Section 4.3 and 4.4*
3 The diagram shows an object of mass 50 kg, which is supported by two cables. Find the tension in each of the supporting cables.	*Section 4.5*
4 The diagram shows a spring fixed to a wall and a block of mass 20 kg, that is at rest on a rough horizontal plane. A string attached to the block passes over a small smooth pulley and is attached to a second block of mass 5 kg. The situation is shown in the diagram. The tension in the spring is 8 N when the block is on the point of sliding towards the pulley. **(a)** Find the coefficient of friction between the block and the plane. **(b)** Describe what happens to the 20 kg block if the string attached to it is cut. Give reasons to support your answer.	*Section 4.7*
5 A particle, of mass 5 kg, is at rest on a slope inclined at an angle of 48° to the horizontal. A force, of magnitude 10 N, that is directed up the slope acts on the particle. **(a)** Find the magnitude of the friction force acting on the particle. **(b)** Find an inequality that the coefficient of friction between the particle and the slope must satisfy.	*Section 4.7*

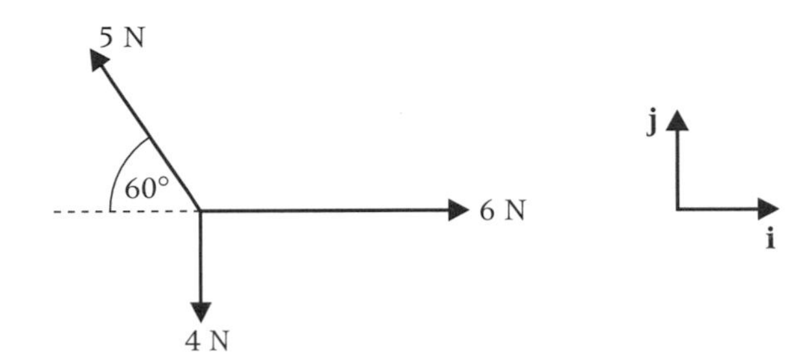

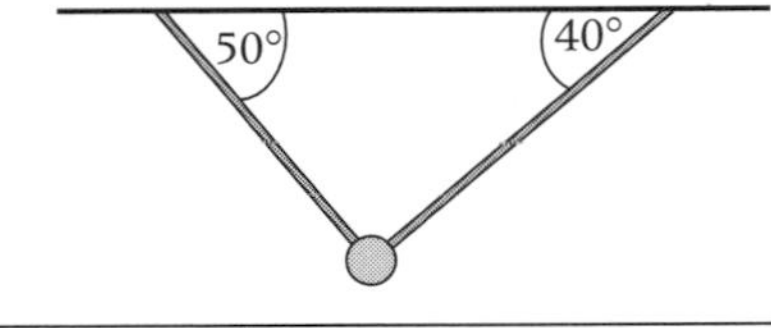

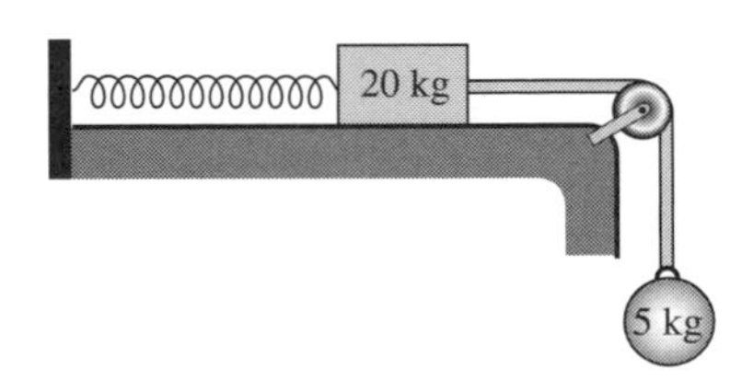

Test yourself ANSWERS

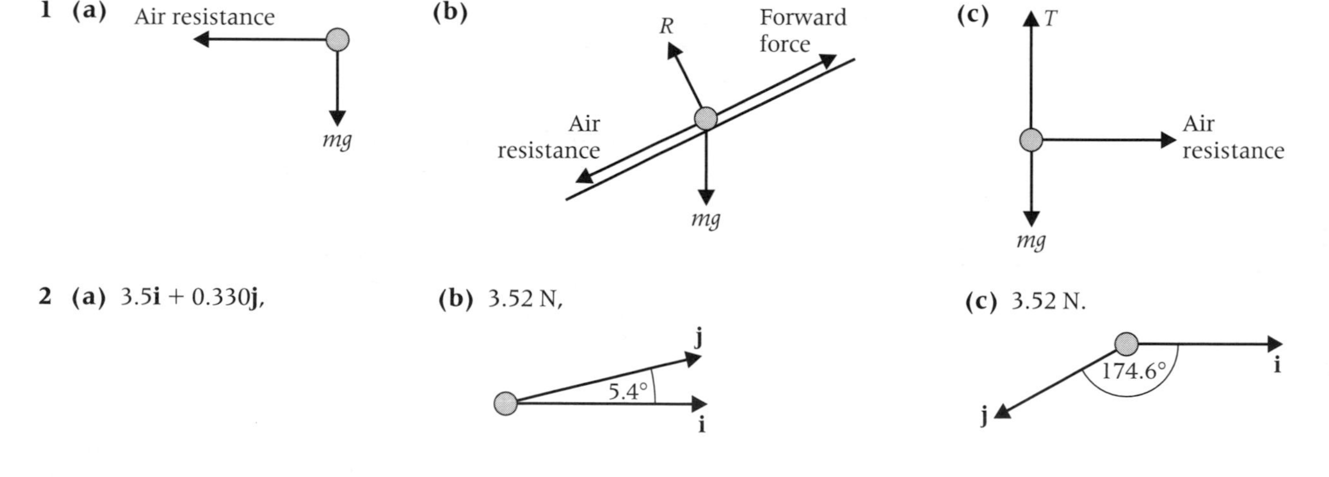

2 **(a)** $3.5\mathbf{i} + 0.330\mathbf{j}$, **(b)** 3.52 N, **(c)** 3.52 N.

3 315 N, 375 N.

4 **(a)** 0.209,

(b) nothing as 8 N is less than $F_{max} = 41$ N.

5 **(a)** 26.4 N,

(b) $\mu \geqslant 0.806$.

CHAPTER 5

Newton's laws of motion

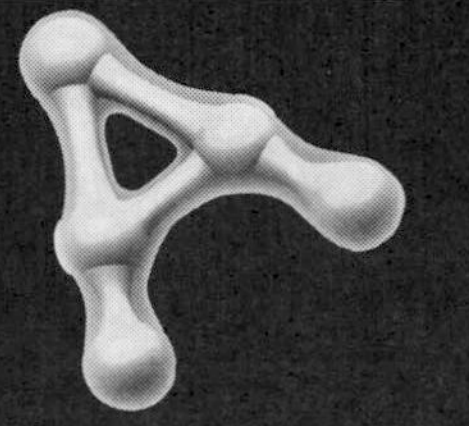

Learning objective

After studying this chapter you should be able to:

- understand the relationship between force and motion as described by Newton's laws.

5.1 Introduction

In this chapter we will examine particles in motion. The relationship between force and motion is described by Newton's laws. In the situations that we will encounter we will use Newton's laws to enable us to model motion produced by forces. If we were to study the motion of atomic particles it would be found that Newton's laws would not provide an accurate enough model and the theory of relativity should be considered.

5.2 Newton's first law

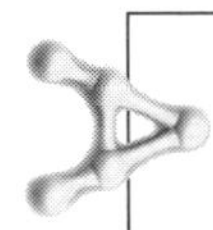

> A body will remain in a state of rest or will continue to move in a straight line with a constant velocity unless it is compelled to change that state by the action of a force.

In the previous chapter we considered the state of equilibrium to be when the resultant force is zero. Newton's first law implies that if a particle moves with a constant velocity then the resultant force will also be zero.

Worked example 5.1

A particle of mass 3 kg slides down a rough plane at an angle α to the horizontal at constant speed. If $\mu = 0.5$ find the angle α.

Solution

The diagram shows the forces acting on the particle.

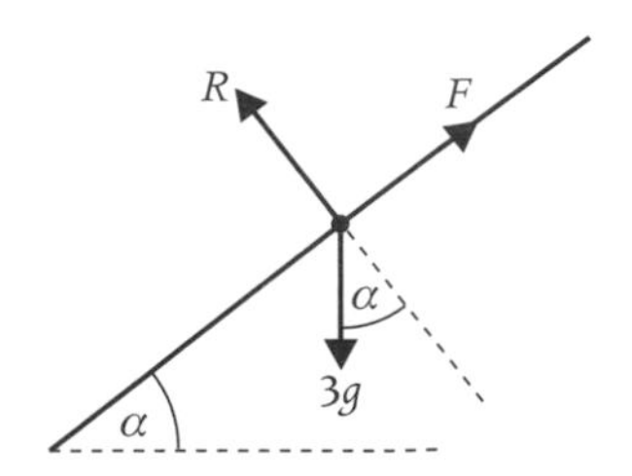

As the particle is moving at a constant speed, the resultant force on the particle is zero.

Resolving horizontally: $mg \sin \alpha = F$

Resolving vertically: $mg \cos \alpha = R$

The laws of friction state that if sliding occurs on a rough surface then the frictional force has reached its maximum and $F = \mu R$, hence

$$F = 0.5R$$

$$mg \sin \alpha = 0.5mg \cos \alpha$$

$$\frac{\sin \alpha}{\cos \alpha} = 0.5$$

$$\tan \alpha = 0.5$$

$$\alpha = 26.6° \text{ (to 3 sf).}$$

Worked example 5.2

A stone of mass 100 g is dropped in a lake. The stone experiences a resistive force which is proportional to the square of its speed, and reaches a maximum speed of 2 m s^{-1}. What will be the maximum speed of a similar stone of mass 50 g?

Solution

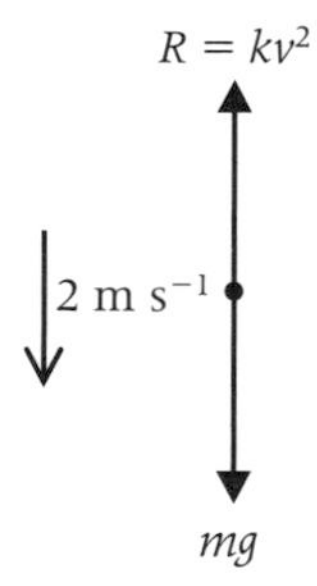

As the stone falls it will increase in speed. The resistive force will therefore increase as well. The maximum speed will be achieved when the resistive force balances the weight of the stone, i.e. $R = mg$.

If R is proportional to v^2 then

$R = kv^2$, where k is a constant.

Hence at maximum speed

$$mg = kv^2$$

$$0.1 \times 9.8 = k \times 2^2$$

$$k = 0.245$$

The resistive force on the similar stone will be a function of its speed and shape, and not influenced by its mass. If we assume it has the same shape as the first stone then we can use $R = 0.245v^2$ again.

When this stone reaches its maximum speed

$$R = mg$$

$$0.05 \times 9.8 = 0.245v^2$$

$$v = 1.41 \text{ m s}^{-1} \text{ (to 3 sf).}$$

EXERCISE 5A

1 The following situations show a body moving with constant velocity in the direction shown, subject to unknown forces of magnitude X and Y. Find X and Y.

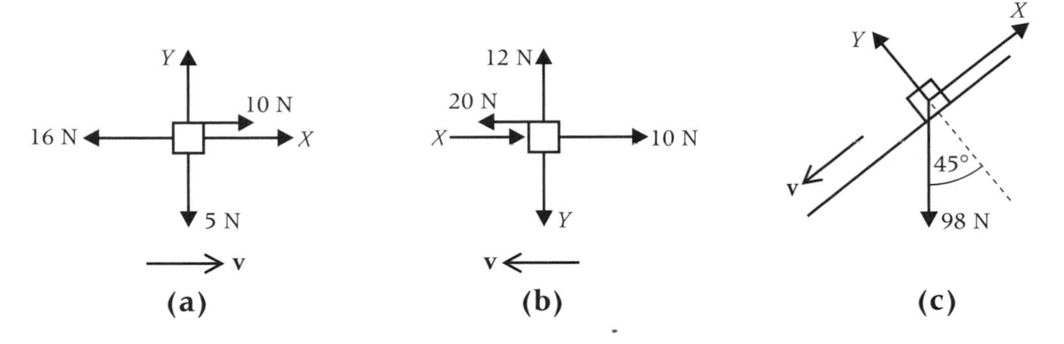

2 The following situations show a body moving with constant velocity in the direction shown, subject to an unknown force of magnitude F N acting an angle θ. Find F and θ.

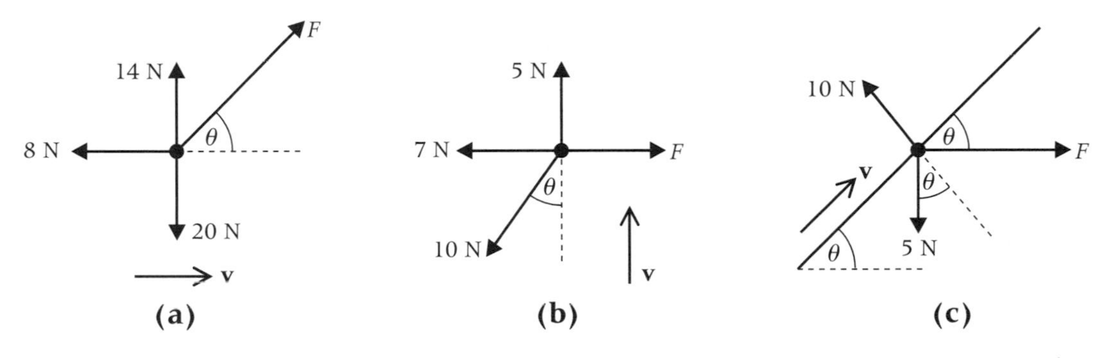

3 A lorry travels along a horizontal road with constant velocity. If the force which the engine exerts is 1350 N, what is the magnitude of the resistive force on the lorry?

4 A small object is being pulled across a horizontal surface at a constant velocity by a force of 12 N acting parallel to the surface. If the mass of the object is 5 kg, determine the coefficient of friction between the object and the surface.

5 A sledge of mass 16 kg is being pulled up the side of a hill of inclination 25°, at a constant velocity. The coefficient of friction between the sledge and the hill is 0.4, and the rope pulling the sledge exerts a force of magnitude T N at 15° to the hill.

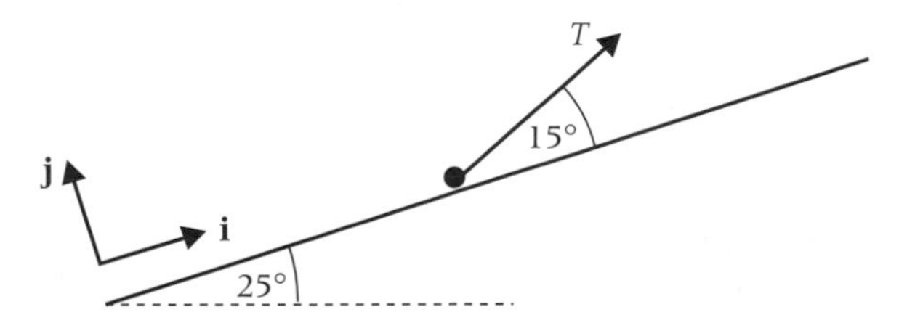

Taking **i** and **j** as unit vectors parallel and perpendicular to the plane, as shown, and modelling the sledge as a particle, write each force as a vector and find:

(a) the magnitude of the tension in the rope, and

(b) the magnitude of the normal reaction between the hill and the sledge.

6 A car of mass 1 tonne travels up an incline of $\sin^{-1}(1/20)$, with constant velocity. If the engine exerts a force of 700 N, what is the magnitude of the resistive force on the car?

7 A lorry of mass 20 tonnes can climb an incline of $\sin^{-1}(1/20)$ at a steady speed of 10 m s^{-1}, when the force produced by the engine is 10.2 kN. Assume that the air resistance is proportional to the square of the speed and ignoring other forms of resistance, find the maximum speed of the lorry when freewheeling down the same hill.

8 A lorry when fully laden weighs 40 tonnes. Its maximum speed when freewheeling down an incline of $\sin^{-1}(1/10)$ is 40 m s^{-1}, subject to air resistance which is proportional to the square of the speed of the lorry. When empty the lorry weighs 8 tonnes. What would be the maximum speed of the empty lorry when freewheeling down the same hill?

9 A skier of mass 80 kg can achieve a maximum speed of 35 m s^{-1} down an incline 30° to the horizontal, subject to: air resistance which is proportional to her speed; and friction where the coefficient of friction between skier and the slope is 0.1. What will be the maximum speed of the skier down an incline of 45° to the horizontal?

5.3 Newton's second law

Newton's second law describes the relationship between the change in motion of a body and the resultant force acting on the body. Provided the mass of the object concerned does not change, as it might if it was a space rocket, then the law can be stated mathematically as:

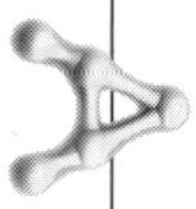

$$F = ma$$

where F is the resultant force on the body, which has mass m and a is its acceleration.

Note that:

(i) F and a are vector quantities. So the acceleration is in the direction of the resultant force.

(ii) The unit of mass is the kg. We measure acceleration in terms of m s^{-2}. A force of one newton (N) is defined as that force necessary to give a mass of 1 kg an acceleration of 1 m s^{-2}.

Worked example 5.3

Find the acceleration produced by a force of 20 N on a mass of 4 kg.

Solution

Using $F = ma$ with $F = 20$ and $m = 4$ gives:

$$20 = 4a$$

$$a = 5\ \text{m s}^{-2}$$

Worked example 5.4

Find the acceleration produced by the forces $(\mathbf{i} + 3\mathbf{j})$ N, $(3\mathbf{i} + \mathbf{j})$ N and $(4\mathbf{i} + 2\mathbf{j})$ N which act on a mass of 2 kg.

Solution

The resultant force is:

$$\mathbf{i} + 3\mathbf{j} + 3\mathbf{i} + \mathbf{j} + 4\mathbf{i} + 2\mathbf{j} = 8\mathbf{i} + 6\mathbf{j}.$$

Using $\quad F = ma$

$$2a = 8\mathbf{i} + 6\mathbf{j}$$

$$a = 4\mathbf{i} + 3\mathbf{j}$$

Note that the magnitude of the acceleration is $\sqrt{4^2 + 3^2} = 5\ \text{m s}^{-2}$.

Worked example 5.5

A particle of mass 3 kg is pulled across a rough horizontal plane, by a light inextensible string inclined at 30° to the horizontal. The tension in the string is 40 N. The coefficient of friction between the particle and the plane is 0.5. Find the acceleration of the particle.

Solution

The diagram shows the forces acting and the acceleration of the particle.

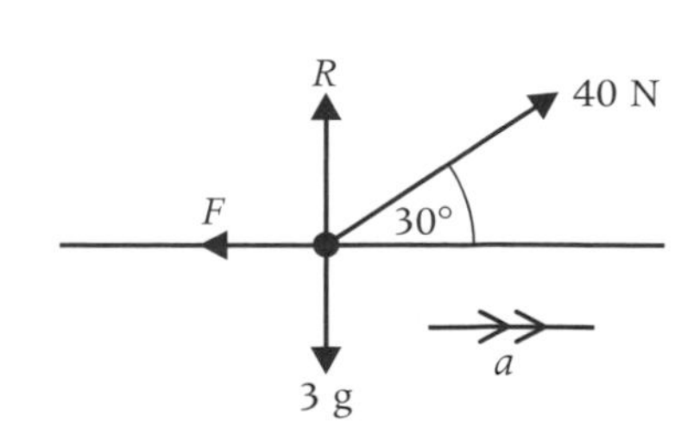

The particle accelerates along the plane so the resultant force must also be parallel to the plane. So the sum of the vertical components of the forces must be zero.

Resolving vertically:

$$R + 40 \sin 30° - 3\,g = 0$$

$$R = 9.4 \text{ N}$$

The frictional force will be at its maximum, i.e.

$$F = \mu R = 0.5 \times 9.4$$

$$F = 4.7 \text{ N}$$

Using $F = ma$ parallel to the plane:

$$40 \cos 30° - 4.7 = 3a$$

$$a = 9.98 \text{ m s}^{-2}$$

Worked example 5.6

A stone is projected with speed 10 m s^{-1} down the line of greatest slope of a plane inclined at 30° to the horizontal. The coefficient of friction between the slope and the stone is 0.8. Find the distance travelled by the stone before coming to rest.

Solution

The diagram shows the forces acting and the acceleration of the particle. The first step is to find the acceleration of the stone.

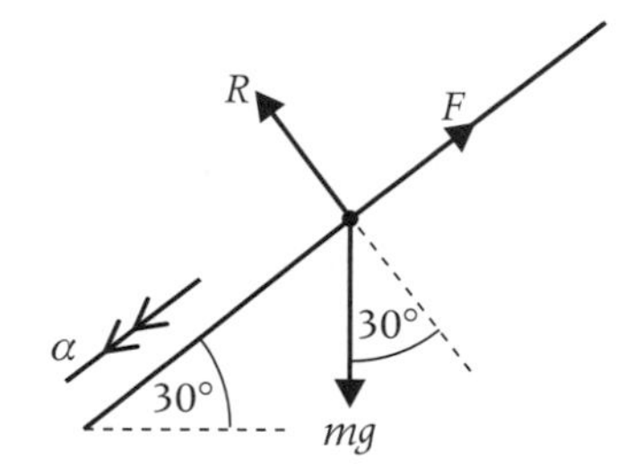

The components of the forces perpendicular to the plane must be in equilibrium. So resolving perpendicular to the plane gives

$$R = mg \cos 30°.$$

Friction is limiting, i.e.

$$F = \mu R$$

$$= 0.8\, mg \cos 30°$$

Using Newton's second law parallel to the plane gives:

$$mg \sin 30° - F = ma$$

$$mg \sin 30° - 0.8\, mg \cos 30° = ma$$

The mass cancels throughout this equation and we get:

$$a = -1.89 \text{ m s}^{-2} \text{ (to 3 sf)}$$

Throughout this motion all the forces are constant so the acceleration will also be constant. If the acceleration is constant we can use the constant acceleration formulae.

Using

$$v^2 = u^2 + 2as$$

gives

$$0^2 = 10^2 + 2 \times (-1.89)d$$

where d is the distance the stone slides. This gives

$$d = 26.5 \text{ m}$$

EXERCISE 5B

1 A single force of magnitude 8 N acts on a particle of mass 16 kg. Find the acceleration produced.

2 The acceleration of a particle of mass 2 kg is 3 m s^{-2}. Find the magnitude of the resultant force on the particle.

3 Forces of $(\mathbf{i} + \mathbf{j})$ N, $(3\mathbf{i} - 2\mathbf{j})$ N and $(\mathbf{i} + 3\mathbf{j})$ N act on a particle of mass 2 kg. Find the acceleration of the particle in the form $a\mathbf{i} + b\mathbf{j}$.

4 A crane lifts a load of mass 2 tonnes vertically from rest. The tension in the cable is 21 kN. Find the acceleration of the load.

5 A package with mass 300 kg is lifted vertically upwards. Find the tension in the cable which lifts the package, when the package:

(a) accelerates upwards at 0.1 m s^{-2},

(b) accelerates downwards at 0.2 m s^{-2},

(c) travels upwards with a retardation of 0.1 m s^{-2}.

6 Find the unknown accelerations, forces and angles in the following situations.

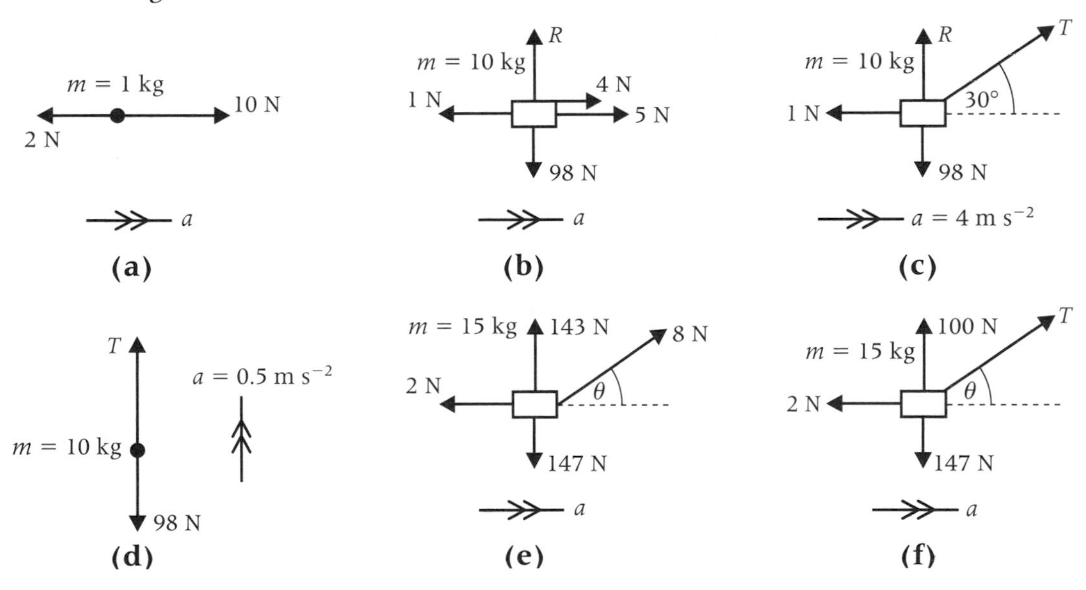

7 Two forces of magnitude 5 N and 6 N act on a particle of mass 2 kg. If the angle between the forces is 60° find the magnitude of the acceleration.

8 Two forces of magnitude 5 N and 6 N act on a particle of mass 2 kg. Find the angle between the forces if the acceleration is 4 m s^{-2}.

9 A car of mass 1 tonne travels along a horizontal road and brakes from 50 m s^{-1} to rest in a distance of 300 m. Find the braking force on the car.

10 A particle of mass m kg slides down a smooth plane, inclined at an angle of 30° to the horizontal. Find the acceleration of the particle down the plane.

11 A particle of mass 6 kg starts from rest and accelerates uniformly. The resultant force on the particle has magnitude 15 N. Find the time taken to reach a speed of 10 m s^{-1}.

12 A particle of mass 20 kg is pulled across a rough horizontal plane by a light inextensible string, inclined at 30° to the horizontal. If the tension in the string is 50 N and the acceleration produced is 0.5 m s^{-2} find the frictional force on the particle and the coefficient of friction.

13 A particle of mass m kg slides down a smooth inclined plane with an acceleration of 2 m s^{-2}. Find the inclination of the plane to the horizontal.

14 A crane lifts a load of 5 tonnes through a vertical height of h metres. The load starts from rest and accelerates uniformly for 2 s to a speed of 0.5 m s^{-1}. It then travels at constant speed for 20 s, after which it is brought to rest in 4 s. Find the tension in the cable during each part of the motion, and the distance travelled by the load.

15 A particle of mass 2 kg starts from rest, and accelerates uniformly, subject to a single force, of magnitude P N, until it reaches a speed 30 m s^{-1}. It then travels with constant velocity for 10 s, after which time a single force of magnitude $2P$ N, acting in the opposite direction to its velocity brings the particle to rest. If the total distance travelled in the motion is 435 m find P.

16 A particle of mass 500 g starts from rest at the point A which has position $(4\mathbf{i} + 2\mathbf{j})$ m. The resultant force on the particle is $(\mathbf{i} + 2\mathbf{j})$ N. Find the position vector of the particle after 3 s.

17 Two tugs are towing a large oil tanker into harbour. Tug A's engines can produce a pulling force of 80 kN while tug B's engines can produce 65 kN of force.

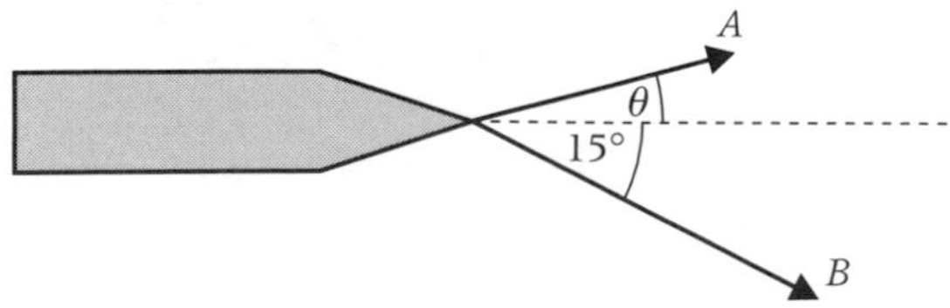

(a) Calculate the angle θ necessary for the tanker to move directly forwards.

(b) Given that there is a resistance to the motion of the tanker of 25 kN directly opposing motion, find the magnitude of the resultant force on the tanker to the nearest 100 N. Find also the acceleration of the tanker if it has a mass of 20 000 tonnes.

18 A sledge of mass 30 kg is accelerating down a hill while a boy is trying to prevent it from sliding by pulling on a rope attached to the sledge with a force of 40 N. The hill has inclination 28° and the rope is inclined to the hill at 10°. The coefficient of friction between the sledge and the hill is 0.35. Find:

(a) the magnitude of the normal contact force on the sledge,

(b) the resultant force on the sledge acting down the hill,

(c) the magnitude of the sledge's acceleration.

19 As a lift moves upwards from rest it accelerates at 0.8 m s^{-2} for 2 s, then travels 4 m at constant speed and finally slows down, with a constant deceleration, stopping in 3 s. The mass of the lift and its occupants is 400 kg.

(a) Find the total distance travelled by the lift and the total time taken.

(b) The lift hangs from a single light, inextensible cable. During which stage of the motion of the lift is the tension in the cable greatest? Find the magnitude of this tension during this stage of the motion. [A]

20 A tree trunk, of mass 250 kg, is pulled up a slope by a chain attached to a tractor. The chain is at an angle of 10° to the slope. The slope itself is at 8° to the horizontal. The tree trunk initially accelerates at 0.2 m s^{-2}. A friction force, of magnitude 2000 N, acts on the tree trunk.

(a) Model the tree trunk as a particle. Draw and label a diagram to show the forces acting on it.

(b) Find the initial tension in the chain.

(c) Explain why the tension in the chain will probably decrease. [A]

21 An amateur inventor is trying to design a simple device that will warn car drivers if their acceleration or deceleration is excessive. A prototype is shown in the diagram, and consists of two strings and a small sphere of mass m attached to the roof of a car at the points A and C.

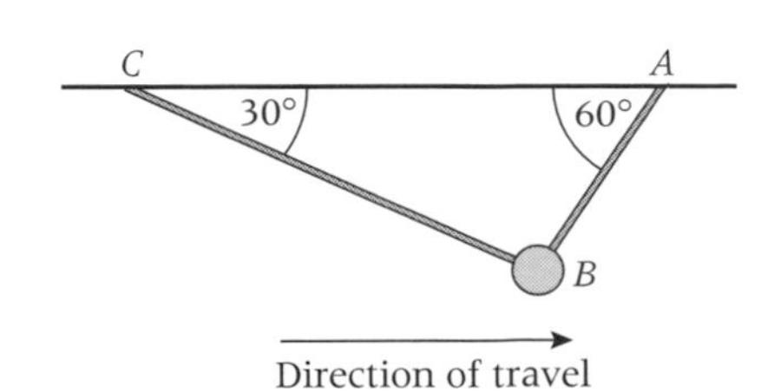

Assume that the car travels in a straight line and that the plane ABC is vertical and parallel to the direction of travel.

Also assume the line AC is horizontal. The acceleration of the car is a.

(a) Show that the tension in the string BC is

$$\frac{m(g - a\sqrt{3})}{2}$$

and find the tension in the other string.

(b) Find the values of a for which each string becomes slack. Is it likely that either string will become slack in normal use? Give reasons for your answer.

(c) Would using a lighter mass reduce the magnitude of the acceleration required for the strings to become slack? Give a reason for your answer. [A]

22 A train is travelling at a constant speed of 40 m s^{-1}, when the driver sees a warning light. Over the next 1000 m the speed of the train drops to 20 m s^{-1}. The train travels at this speed for 5 minutes. The speed returns to 40 m s^{-1} after a further 5 minutes. Assume that the acceleration of the train is constant on each stage of its journey.

(a) Find the total distance travelled by the train, while its speed is less than its normal operating speed of 40 m s^{-1}.

(b) The train would normally have travelled this distance at a constant 40 m s^{-1}. Find the time by which it was delayed.

(c) The train has a mass of 50 tonnes and moves on a straight, horizontal track. Assume that while the train is slowing down, it experiences a constant resistive force of 40 000 N and one other force parallel to the track. Find the magnitude of this force and state the direction in which it acts. [A]

23 A child is sliding at a constant speed of 4 m s^{-1} down a long slide. The child has a mass of 45 kg. The slide is inclined at an angle of 40° to the horizontal. Assume that a constant friction force, of magnitude 89 N, acts on the child.

(a) Use the data given to explain why air resistance must be taken into account when modelling the motion of the child. Find the magnitude of the air resistance acting on the child, when he is travelling at a constant speed of 4 m s^{-1}.

(b) Assume that the magnitude of the air resistance is proportional to the speed of the child. The next time that he uses the slide he starts from rest and accelerates. Find his acceleration, when he is moving at 1 m s^{-1}. [A]

5.4 Newton's third law

Newton's third law states:

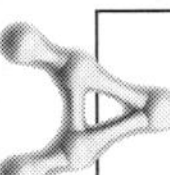

For every action, there is an equal but opposite reaction.

This law is often misunderstood, but is really very simple. The best way to understand it is to consider some simple examples.

If you stand on a table, your feet push down on the table. In response to this the table exerts an upward force on your feet. This upward force is the equal but opposite reaction to the downward force you exert on the table. These are both normal reaction forces.

The diagrams show these normal reaction forces (R).

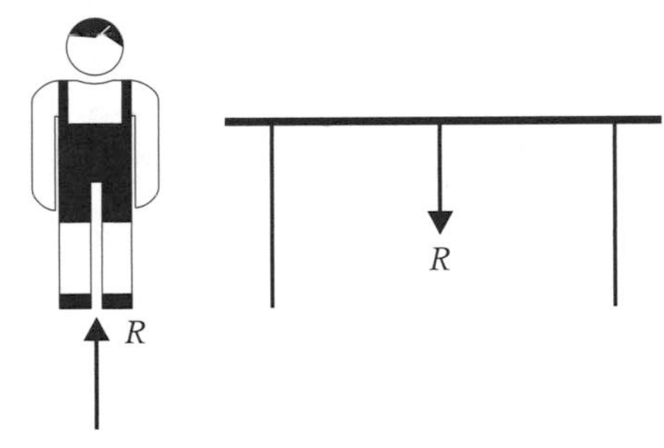

It is important to note that this example does not directly involve the weight of the person (W), although $W = R$.

A second example concerns the orbit of the moon around the earth. The earth exerts a gravitational force on the moon and the moon exerts an equal but opposite force on the earth. The effect of the force acting on the moon is to keep it in orbit around the earth. The effect of the force on the earth is to cause the tides. The important idea, however, is that both planets exert forces of equal size on each other, as shown in the diagram.

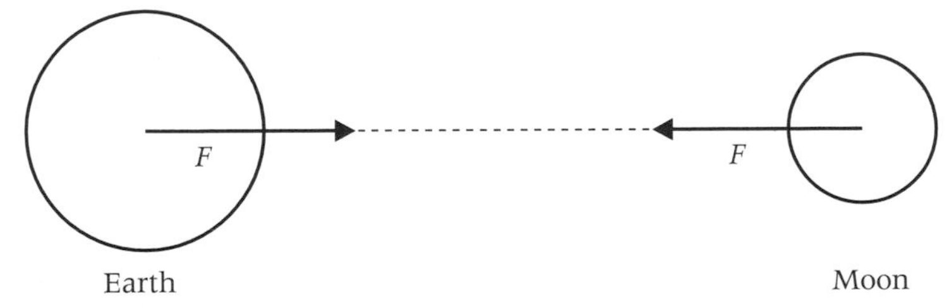

As a final example consider a car towing a caravan. The car exerts a forward force on the caravan, but the reaction to this is the backwards force that the caravan exerts on the car.

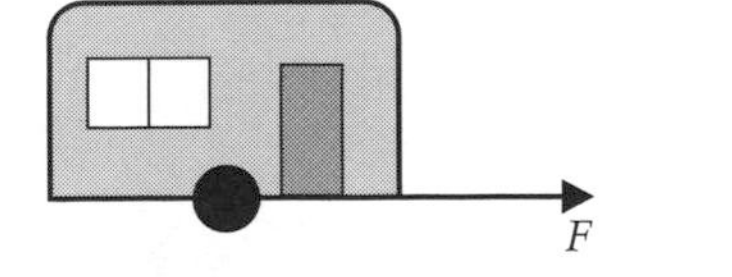

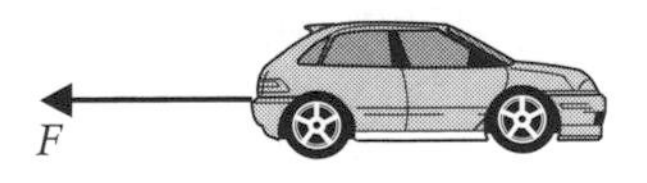

EXERCISE 5C

1 Use Newton's third law to explain why you would hurt your hand if you punched a hard object with it.

2 A ladder leans against a smooth wall.

(a) Draw diagrams to show the force that the top of the ladder exerts on the wall and the force that the wall exerts on the top of the ladder.

(b) Use Newton's third law to explain why these forces have equal magnitudes.

3 A child jumps off a table and lands on the ground. Describe how the force that the ground exerts on the child varies. Also describe how the force that the child exerts on the ground varies.

4 A man, of mass 78 kg, stands in a lift of mass 200 kg that is accelerating upwards at 0.5 m s^{-2}. Calculate the magnitudes of the forces that act on the lift. Also draw a diagram to show how they act.

5 Three carriages are coupled to an engine on a set of railway lines. The carriages and engine move forward on horizontal tracks. Draw diagrams to show the forces acting on each of the carriages and the trucks. Clearly show any forces that have the same magnitude.

Key point summary

Formula to learn

$F = ma$

1 If a body is at rest or moves with a constant velocity the forces acting on it must be in equilibrium. p73

2 When applying Newton's second law, remember that F represents the resultant force. p76

3 When two bodies interact, the force exerted by the first body on the second body is equal and opposite to the force exerted by the second body on the first. p83

Test yourself	What to review
1 A helicopter of mass 880 kg is rising vertically at a constant rate. Find the magnitude of the lift force acting on the helicopter. How would your answer change if the helicopter was descending at a constant rate?	*Section 5.2*
2 A child, of mass 30 kg, slides down a slide at a constant speed. Assume that there is no air resistance acting on the child. The slide makes an angle of 40° with the horizontal. Find the magnitude of the friction force on the child and the coefficient of friction.	*Section 5.2*
3 A lift and its passengers have a total mass of 300 kg. Find the tension in the lift cable if: **(a)** it accelerates upwards at 0.2 m s^{-2}, **(b)** it accelerates downwards at 0.05 m s^{-2}.	*Section 5.3*
4 A van, of mass 1200 kg, rolls down a slope, inclined at 3° to the horizontal and experiences a resistance force of magnitude 400 N. Find the acceleration of the van.	*Section 5.3*

Test yourself ANSWERS

1 8624 N, no change.

2 189 N, 0.839.

3 **(a)** 3000 N, **(b)** 2925 N.

4 0.180 m s^{-2}.

CHAPTER 6

Kinematics and variable acceleration

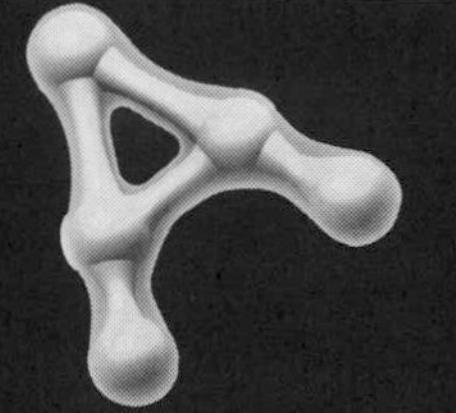

Learning objectives

After studying this chapter you should be able to:

- differentiate displacements or position vectors to give velocities and accelerations for one, two or three dimensions
- integrate accelerations to give velocities and position vectors or displacements.

6.1 Introduction

In Chapters 2 and 3 we have considered only cases where the acceleration of an object is constant. For example a ball falling under gravity or perhaps a car with a constant acceleration. However there are many situations where it is unrealistic to assume that the acceleration of a body is constant, and where better solutions to problems can be obtained by modelling the acceleration as variable. For example circular motion involves an acceleration that is always changing direction and the acceleration of a car may decrease as it gains speed.

In some cases the acceleration can be expressed as a function of time, in others it may depend on the speed or velocity of the object under consideration. In this chapter you will consider cases where the acceleration is dependent on time, but in later chapters you will consider cases where the acceleration depends on displacement or velocity.

This chapter will require you to use calculus instead of the constant acceleration equations, from now on when approaching a problem it is important to decide whether or not the acceleration is constant before using the constant acceleration equations.

6.2 Displacement to velocity and acceleration

In Chapter 2 it was noted that the velocity was given by the gradient of a displacement–time graph and that the acceleration was given by the gradient of a velocity–time graph. These results can be used as the basis of your work with variable acceleration.

The gradient of a curve is given by its derivative so we can deduce that the velocity is given by the derivative, with respect to time, of the displacement.

$$v = \frac{dx}{dt}$$

The velocity is equal to the rate of change of the displacement.

Similarly the acceleration will be given by the derivative of the velocity, with respect to time.

$$a = \frac{dv}{dt}$$

The acceleration is equal to the rate of change of the velocity.

You can also write

$$a = \frac{d^2x}{dt^2}.$$

The following worked examples illustrate how these results can be applied.

Worked example 6.1

The height of a bullet, h metres, fired vertically upwards, at time t seconds, is given by:

$$h = 3 + 80t - 4.9t^2$$

Show that the acceleration of the bullet is constant and find its maximum height.

Solution

First differentiate h to find the velocity.

$$v = \frac{dh}{dt}$$

$$= 80 - 2 \times 4.9t$$

$$= 80 - 9.8t$$

6

Now differentiate again to find the acceleration.

$$a = \frac{dv}{dt}$$
$$= -9.8$$

So the acceleration is constant and has magnitude 9.8 m s^{-2}.

The maximum height will be attained when the velocity is zero.

$$80 - 9.8t = 0$$
$$t = \frac{80}{9.8}$$

This can then be substituted into the expression for h to find the maximum height.

$$h = 3 + 80 \times \frac{80}{9.8} - 4.9 \times \left(\frac{80}{9.8}\right)^2$$
$$= 330 \text{ m (to 3 sf)}$$

This example shows that calculus can also be applied to cases involving constant acceleration.

Worked example 6.2

As a car slows down the distance, s metres, it has travelled at time, t s, is modelled by the equation:

$$s = \frac{t^4}{8} - t^3 + 16t + 32 \quad \text{for } 0 \leqslant t \leqslant 4.$$

(a) Show that when $t = 4$ the car has zero velocity.

(b) Find the acceleration, when $t = 0$, $t = 2$ and $t = 4$.

(c) Describe how the resultant force on the car changes.

Solution

(a) First you need to differentiate s to find an expression for the velocity.

$$v = \frac{ds}{dt}$$
$$= \frac{4t^3}{8} - 3t^2 + 16$$
$$= \frac{t^3}{2} - 3t^2 + 16$$

Now substitute $t = 4$ in this expression.

$$v = \frac{4^3}{2} - 3 \times 4^2 + 16$$
$$= 32 - 48 + 16$$
$$= 0$$

(b) Now differentiate again to find the acceleration.

$$a = \frac{dv}{dt}$$
$$= \frac{3t^2}{2} - 6t$$

Substituting the values $t = 0$, $t = 2$ and $t = 4$ gives

$$t = 0, a = \frac{3 \times 0^2}{2} - 6 \times 0 = 0 \text{ m s}^{-2}$$
$$t = 2, a = \frac{3 \times 2^2}{2} - 6 \times 2 = -6 \text{ m s}^{-2}$$
$$t = 4, a = \frac{3 \times 4^2}{2} - 6 \times 4 = 0 \text{ m s}^{-2}$$

(c) The acceleration is always negative, and varies as shown in the graph. Its magnitude increases from 0 m s^{-2} to 6 m s^{-2} and then decreases back to 0 m s^{-2} as the car comes to rest after 4 s.

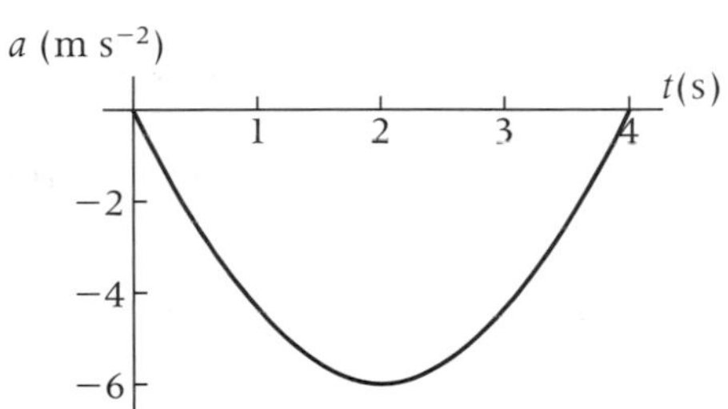

Worked example 6.3

A ball is released from rest at the top of a tall building and falls vertically. The distance fallen by the ball at time t s is x m where

$$x = 5t + 2.5e^{-2t} - 2.5$$

(a) Find an expression for the velocity and sketch a graph to show how the velocity varies with time.

(b) Find an expression for the acceleration and sketch a graph to show how this varies with time.

Solution

(a) Differentiate x with respect to t to find v.

$$v = \frac{dx}{dt}$$
$$= 5 - 5e^{-2t}$$

Initially v is 0, but increases to 5 m s^{-1} as t increases, as shown in the graph.

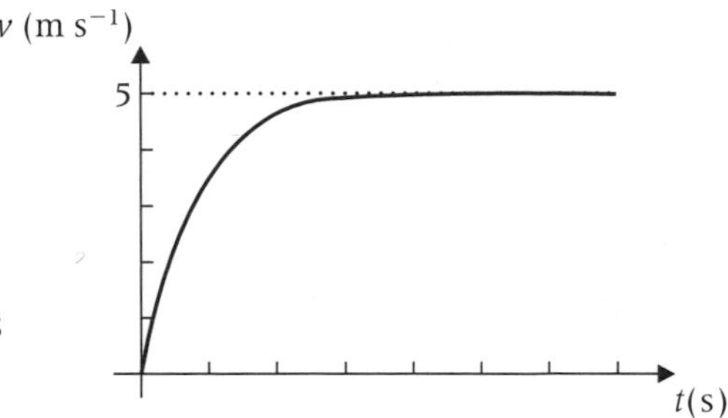

(b) Differentiating again gives the acceleration.

$$a = \frac{dv}{dt}$$
$$= 10e^{-2t}$$

Initially the acceleration is 10 m s^{-2}, but this decreases to 0 as shown in the graph.

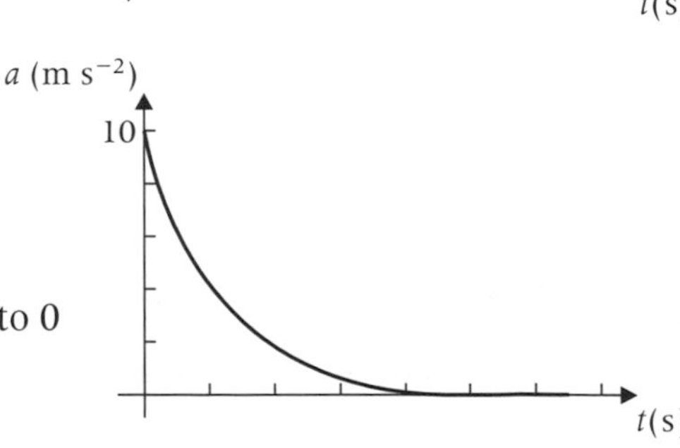

EXERCISE 6A

1 The distances, in metres, travelled by a cyclist after t s are given by

$$s = \frac{t^3}{6} - \frac{t^4}{120} \quad \text{for } 0 \leqslant t \leqslant 10.$$

(a) How far has the cyclist travelled when $t = 10$?

(b) Find an expression for the velocity of the cyclist at time t.

(c) Find an expression for the acceleration of the cyclist at time t.

(d) Describe how the acceleration of the cyclist changes.

2 A car accelerates from rest so that the distance that it has travelled in t s is s m where $s = t^2 - \dfrac{t^3}{60}$.

(a) Find expressions for the velocity and acceleration at time t s.

The expression for s is valid while the acceleration is greater than or equal to zero.

(b) Find the time when the acceleration becomes zero.

(c) Find the velocity of the car when the acceleration is zero.

(d) How far does the car travel before the acceleration becomes zero?

3 A lift rises from ground level. The height, s m, of the lift at time t s is given by $s = \dfrac{3t^2}{10} - \dfrac{t^3}{50}$ for $0 \leqslant t \leqslant 10$.

(a) Describe what happens to the lift when $t = 10$ s.

(b) Sketch a graph to show how the acceleration of the lift varies with time.

4 A car driver sees a red traffic light in front of him and starts to brake. The distance, s m, travelled while the car has been braking for t s is given by

$$s = \frac{45t}{2} - \frac{3t^2}{2} + \frac{t^3}{30}$$

This expression only applies while the car is moving.

(a) Find the range of values of t for which the expression for s is valid.

(b) Find the distance travelled while the car comes to rest.

(c) Sketch an acceleration–time graph for the car.

5 A firework manufacturer is designing a new type of firework. They want it to rise so that the height, h m, at time t s is given by $h = 9t^2 - \frac{t^4}{12}$. The firework should explode when it reaches its maximum height.

Find the maximum height of the firework.

6 A weight is suspended from an elastic string. It moves up and down, so that at time t s the distance between the weight and the fixed end of the string is x m, where $x = 0.8 + 0.4\sin(0.5t)$.

(a) Find the velocity of the weight at time t.

(b) What is the maximum speed of the weight?

(c) Find the acceleration of the weight when $t = 2$.

(d) Find the range of values of the acceleration.

7 The height, h m, of a hot air balloon at time t s is modelled by

$$h = 150\left(1 - \cos\left(\frac{t}{800}\right)\right)$$

The model is only valid while the balloon is gaining height.

(a) State the initial height of the balloon.

(b) Find the range of values of t for which the model is valid.

(c) What is the maximum height of the balloon?

(d) What is the maximum acceleration of the balloon?

8 A particle is set in motion with an initial speed of $20\ \text{m s}^{-1}$ on a smooth horizontal surface. It slows down due to the action of air resistance, stopping after it has travelled 15 m. A possible model for the displacement, s m, at time t s is $s = A(1 - e^{-kt})$, where A and k are constants.

(a) State the value of A.

(b) Find k.

(c) Sketch a graph to show how the acceleration varies with time.

9 A particle, that hangs on a spring, moves so that the displacement, x m, from its equilibrium position at time t s is given by $x = 4\cos 2t + 3\sin 2t$.

(a) Find the initial displacement of the particle.

(b) Find the initial speed of the particle.

(c) Show that the acceleration $a\ \text{m s}^{-2}$, satisfies the relationship $a = -4x$.

10 An object falls through a fluid so that the distance fallen, in metres, at time t s is given by $s = 40(4e^{-\frac{t}{4}} + t - 4)$.

(a) Find the initial and terminal speeds of the object.

(b) Sketch a graph to show how the acceleration of the object varies with time.

11 A particle is projected vertically, so that it moves under the influence of gravity and is subject to air resistance. The height, h m, of the particle at time t s is given by

$$h = \frac{1}{k}\left(\frac{g}{k} + U\right)(1 - e^{-kt}) - \frac{gt}{k}$$

where k and U are constants. The model is only valid while the particle is moving upwards.

(a) Show that the model is valid while

$$0 \leqslant t \leqslant \frac{1}{k}\ln\left(1 + \frac{kU}{g}\right)$$

(b) Find the initial acceleration of the particle and sketch an acceleration–time graph for the particle.

12 A rocket that is launched at a firework display is to be modelled as a particle. The height, h m, of the rocket at t s after lift-off is modelled by

$$h = \frac{5t^2}{2} - \frac{t^4}{20}$$

The rocket rises vertically from rest and this model applies until the speed of the rocket drops to zero, when all the rocket's fuel has been used. The rocket then falls back to the ground.

(a) Find expressions for the velocity and acceleration of the rocket as it rises.

(b) Find the range of values of t for which the above model applies and the maximum height of the rocket in this period of time.

(c) Describe what happens to the acceleration of the rocket while it is rising and find its maximum speed.

(d) State **two** assumptions that it would be appropriate to make about the motion of the rocket after it has reached its maximum height, in order to predict the time it takes to fall back to the ground. [A]

6.3 Acceleration to velocity and displacement

The process of differentiating to move from displacement to velocity and from velocity to acceleration can be reversed using integration.

You can integrate an acceleration to obtain a velocity and integrate a velocity to obtain a displacement.

$$v = \int a dt$$

$$s = \int v dt$$

When using integration in this way it is very important to remember to include the constants of integration, which will depend on the initial velocities and positions of the objects that are under consideration.

The following examples illustrate the use of integration in this context.

Worked example 6.4

A car slows down from a speed of 30 m s^{-1}. Its acceleration, a m s^{-2} at time t s is given by $a = -\frac{t}{2}$. This expression is valid while the car is moving.

(a) Find an expression for the velocity of the car at time t.

(b) Find an expression for the distance travelled by the car at time t.

(c) Find the distance that the car travels before it stops.

Solution

(a) First integrate the acceleration to obtain the velocity, v m s^{-1}.

$$v = \int -\frac{t}{2} dt$$

$$= -\frac{t^2}{4} + c$$

The fact that the initial speed was 30 m s^{-1} can now be used to find c. Substituting $t = 0$ and $v = 30$ gives

$$30 = -\frac{0^2}{4} + c$$

$$c = 30$$

So the velocity at time t is

$$v = 30 - \frac{t^2}{4}$$

(b) The velocity can now be integrated to obtain the displacement, s m.

$$s = \int 30 - \frac{t^2}{4} dt$$

$$= 30t - \frac{t^3}{12} + C$$

If we assume that the car starts at the origin, we can substitute $t = 0$ and $s = 0$, to determine the value of C.

$$0 = 30 \times 0 - \frac{0^3}{12} + C$$

$$C = 0$$

So the displacement at time t is given by

$$s = 30t - \frac{t^3}{12}$$

(c) The first step is to find t when the car stops.

$$0 = 30 - \frac{t^2}{4}$$

$$t^2 = 120$$

$$t = \sqrt{120}$$

This value for t can now be substituted into the expression obtained for the displacement.

$$s = 30 \times \sqrt{120} - \frac{(\sqrt{120})^3}{12}$$

$$= 219 \text{ m (to 3 sf)}$$

Worked example 6.5

As a cyclist sets off from rest the acceleration, a m s^{-2}, of the cyclist, at time t s is given by $a = 1 - \frac{t}{20}$ for $0 \leqslant t \leqslant 20$.

(a) Find expressions for the velocity and displacement of the cyclist at time t.

(b) What is the speed of the cyclist after 20 s?

(c) How far does the cyclist travel in the 20 s?

Solution

(a) The acceleration should be integrated to give the velocity.

$$v = \int 1 - \frac{t}{20} dt$$

$$= t - \frac{t^2}{40} + c$$

As the cyclist starts at rest, substituting $t = 0$ and $v = 0$ will give c.

$$0 = 0 - \frac{0^2}{40} + c$$

$$c = 0$$

So the velocity at time t is given by $v = t - \dfrac{t^2}{40}$.

This can now be integrated to give the displacement.

$$s = \int t - \frac{t^2}{40} dt$$

$$= \frac{t^2}{2} - \frac{t^3}{120} + C$$

If we assume that the cyclist starts at the origin, then substituting $t = 0$ and $x = 0$ will give the value of C.

$$0 = \frac{0^2}{2} - \frac{0^3}{120} + C$$

$$C = 0$$

So the displacement at time t is given by $s = \dfrac{t^2}{2} - \dfrac{t^3}{120}$.

(b) We can now substitute $t = 20$ into the expression for the velocity.

$$v = 20 - \frac{20^2}{40}$$

$$= 10 \text{ m s}^{-1}$$

The cyclist reaches a speed of 10 m s^{-1}.

(c) Substituting $t = 20$ into the expression for s will give the distance travelled.

$$s = \frac{20^2}{2} - \frac{20^3}{120}$$

$$= 133\tfrac{1}{3} \text{ m}$$

Worked example 6.6

A parachutist is initially falling at a constant speed of 30 m s^{-1}, when he opens his parachute The acceleration, a m s^{-2}, of the parachutist t s after the parachute has been opened is modelled as $a = -13e^{-\frac{t}{2}}$.

(a) Find expressions for the velocity and the distance fallen by the parachutist at time t.

(b) Sketch a velocity–time graph for the parachutist.

(c) The parachutist must have a speed of less than 5 m s^{-1} when he lands. Find the minimum height at which he can open his parachute.

Solution

(a) First integrate the acceleration to find the velocity.

$$v = \int -13e^{-\frac{t}{2}} dt$$

$$= 26e^{-\frac{t}{2}} + c$$

Using the initial velocity, you can substitute $v = 30$ and $t = 0$ to find c.

$$30 = 26e^0 + c$$

$$c = 4$$

So the velocity at time t is given by $v = 26e^{-\frac{t}{2}} + 4$.

This can now be integrated to give the distance that has been fallen.

$$s = \int 26e^{-\frac{t}{2}} + 4dt$$

$$= -52e^{-\frac{t}{2}} + 4t + C$$

As the distance fallen will initially be zero you can substitute $t = 0$ and $s = 0$ to find C.

$$0 = -52e^0 + 4 \times 0 + C$$

$$C = 52$$

The distance fallen at time t is then

$$s = 52 - 52e^{-\frac{t}{2}} + 4t$$

$$= 52(1 - e^{-\frac{t}{2}}) + 4t$$

(b) The expression for the velocity shows that it decreases exponentially from 30 m s^{-1} to 4 m s^{-1}. This is shown in the graph.

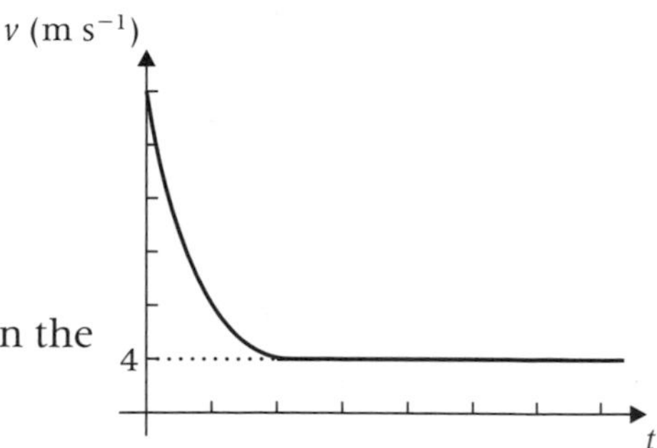

(c) To find when the speed of the parachutist drops to 5 m s^{-1}, we must solve the equation below.

$$\begin{aligned} 5 &= 26e^{-\frac{t}{2}} + 4 \\ e^{-\frac{t}{2}} &= \frac{1}{26} \\ -\frac{t}{2} &= \ln\left(\frac{1}{26}\right) \\ t &= 2\ln(26) \\ &= 6.52 \text{ s} \end{aligned}$$

Now this value of t can be substituted into the expression for s.

$$\begin{aligned} s &= 52\left(1 - e^{-\frac{2\ln 26}{2}}\right) + 4 \times 2\ln 26 \\ &= 76.1 \text{ m (to 3 sf)} \end{aligned}$$

So the minimum safe height to open the parachute is 76.1 m.

EXERCISE 6B

1 The acceleration, $a \text{ m s}^{-2}$, at time t s of a particle that starts at rest is given by $a = \frac{t^2}{100}$.

(a) Find an expression for the velocity of the particle at time t.

(b) Find the velocity of the particle when $t = 5$.

(c) Find the distance that the particle travels in the first 5 seconds of its motion.

2 The acceleration of a cyclist, at time t s is given by $a = 2 - \frac{t}{10} \text{ m s}^{-2}$. This model is valid until $t = 20$ s. If the cyclist starts at rest, find the distance travelled in the 20 s and the final speed of the cyclist.

3 A body experiences an acceleration of $0.1t \text{ m s}^{-2}$, at time t s, for $0 \leqslant t \leqslant 5$. Its acceleration is zero for $t > 5$. Find the distance travelled by the body when $t = 10$, if it has an initial velocity of 3 m s^{-1}.

4 A train travels along a straight set of tracks. Initially it moves with velocity 10 m s^{-1}. It then experiences an acceleration given by $a = 0.1t(4 - t) \text{ m s}^{-2}$, until $t = 4$ s. Find the velocity when $t = 4$ and the distance travelled in this time.

5 The graph below shows how the magnitude of the force exerted on a lorry, of mass 10 000 kg, by its brakes varies with time. The lorry initially has a velocity of 20 m s^{-1}.

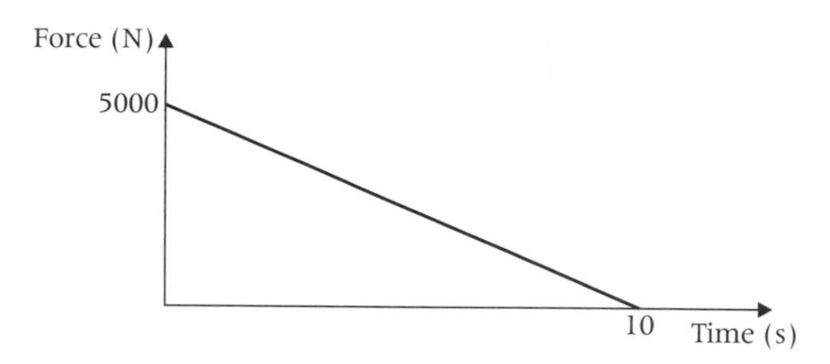

(a) Show that the acceleration, a m s^{-2}, at time t s is given by $a = \frac{t}{20} - \frac{1}{2}$ for $0 \leqslant t \leqslant 10$.

(b) Find the velocity of the lorry when $t = 10$.

(c) Find the distance travelled by the lorry in the 10 s period.

6 The acceleration of a car at time t s is $\frac{20 - t}{10}$ m s^{-2}.

The car starts at rest.

(a) Find an expression for the velocity of the car at time t.

(b) When is the acceleration of the car zero and what is its speed at this time?

(c) The formula given above applies for $0 \leqslant t \leqslant 30$. Describe how the car moves during the 30 s.

(d) Find the total distance travelled by the car in the 30 s period.

7 A mass that is attached to one end of a spring moves up and down. The velocity, v m s^{-1}, at time t s of the mass is given by $v = 0.6 \sin 3t$.

(a) Find the acceleration of the mass at time t.

(b) Find the displacement of the mass at time t if its initial displacement is 0.8 m from the fixed end of the spring.

(c) Sketch a displacement–time graph for the mass.

8 The acceleration, a m s^{-2}, at time t s of a stone that falls from rest is modelled as $9.8e^{-2t}$.

(a) Find expressions for the velocity and displacement of the stone at time t.

(b) Sketch a velocity–time graph for the stone.

(c) How far would the stone fall in 20 s?

9 A bullet is fired vertically upwards, from ground level, with an initial velocity of 40 m s^{-1}. Its acceleration at time t s is modelled as $-50e^{-0.5t}$ m s^{-2}. This model only applies while the bullet is rising.
Find the maximum height of the bullet.

10 The tip of the blade in an electric jigsaw moves so that its acceleration is $40\cos(100\pi t)$ m s^{-2}, t s after it starts to move from rest at its lowest position.

(a) Find the velocity of the tip of the blade at time t.

(b) What is the maximum velocity of the tip of the blade?

(c) What is the maximum displacement of the tip of the blade from its lowest position?

11 A particle moves so that its acceleration at time t s is $4 \sin 2t + 6 \cos 2t$ m s^{-2}. Initially the particle has a velocity of -2 m s^{-1} and is at the origin.

(a) Find an expression for the position of the particle at time t.

(b) Determine the maximum speed of the particle.

6.4 Motion in two or three dimensions

The ideas that you have used in one dimension can be very easily extended to two and three dimensions. A position vector would be written as $\mathbf{r} = x\mathbf{i} + x\mathbf{j} + z\mathbf{k}$ in three dimensions or as $\mathbf{r} = x\mathbf{i} + y\mathbf{j}$ in two dimensions. Here x, y and z are all functions of time. To obtain velocities and accelerations we simply need to differentiate these with respect to time. So in three dimensions we have

$$\mathbf{r} = x\mathbf{i} + y\mathbf{j} + z\mathbf{k}$$

$$\mathbf{v} = \frac{d\mathbf{r}}{dt}$$

$$= \frac{dx}{dt}\mathbf{i} + \frac{dy}{dt}\mathbf{j} + \frac{dz}{dt}\mathbf{k}$$

$$\mathbf{a} = \frac{d\mathbf{v}}{dt}$$

$$= \frac{d^2x}{dt^2}\mathbf{i} + \frac{d^2y}{dt^2}\mathbf{j} + \frac{d^2z}{dt^2}\mathbf{k}$$

In two dimensions these results reduce to

$$\mathbf{r} = x\mathbf{i} + y\mathbf{j}$$

$$\mathbf{v} = \frac{d\mathbf{r}}{dt}$$

$$= \frac{dx}{dt}\mathbf{i} + \frac{dy}{dt}\mathbf{j}$$

$$\mathbf{a} = \frac{d\mathbf{v}}{dt}$$

$$= \frac{d^2x}{dt^2}\mathbf{i} + \frac{d^2y}{dt^2}\mathbf{j}$$

These results are used in the following examples.

Worked example 6.7

The position vector, $\mathbf{r}$ m, at time t s of an aeroplane that is circling an airport is given by

$$\mathbf{r} = 500\sin\left(\frac{t}{20}\right)\mathbf{i} + 500\cos\left(\frac{t}{20}\right)\mathbf{j} + 4000\mathbf{k}$$

The unit vectors $\mathbf{i}$ and $\mathbf{j}$ are east and north, respectively, and $\mathbf{k}$ is vertical.

(a) Find the velocity of the aeroplane.

(b) Find the magnitude of the acceleration of the aeroplane.

Solution

(a) The velocity can be found by differentiating the position vector with respect to time.

$$\mathbf{v} = \frac{d\mathbf{r}}{dt}$$

$$= \frac{d}{dt}\left(500\sin\left(\frac{t}{20}\right)\right)\mathbf{i} + \frac{d}{dt}\left(500\cos\left(\frac{t}{20}\right)\right)\mathbf{j} + \frac{d}{dt}(4000)\mathbf{k}$$

$$= 25\cos\left(\frac{t}{20}\right)\mathbf{i} - 25\sin\left(\frac{t}{20}\right)\mathbf{j}$$

(b) The velocity can be differentiated to obtain the acceleration.

$$\mathbf{a} = \frac{d\mathbf{v}}{dt}$$

$$= \frac{d}{dt}\left(25\cos\left(\frac{t}{20}\right)\right)\mathbf{i} + \frac{d}{dt}\left(-25\sin\left(\frac{t}{20}\right)\right)\mathbf{j}$$

$$= -\frac{5}{4}\sin\left(\frac{t}{20}\right)\mathbf{i} - \frac{5}{4}\cos\left(\frac{t}{20}\right)\mathbf{j}$$

Now consider the magnitude of the acceleration.

$$a = \sqrt{\left(-\frac{5}{4}\sin\left(\frac{t}{20}\right)\right)^2 + \left(-\frac{5}{4}\cos\left(\frac{t}{20}\right)\right)^2}$$

$$= \frac{5}{4}\sqrt{\sin^2\left(\frac{t}{20}\right) + \cos^2\left(\frac{t}{20}\right)}$$

$$= \frac{5}{4}\text{ m s}^{-2}$$

Obtaining a position vector

In the same way that accelerations and velocities were integrated in one dimension, you can integrate velocity or acceleration vectors to obtain velocities or displacements. In three dimensions we would integrate the acceleration to obtain the velocity.

$$\mathbf{a} = a_x\mathbf{i} + a_y\mathbf{j} + a_z\mathbf{k}$$

$$\begin{aligned}\mathbf{v} &= \int \mathbf{a}dt \\ &= \int a_x dt\mathbf{i} + \int a_y dt\mathbf{j} + \int a_z dt\mathbf{k}\end{aligned}$$

Similarly we would integrate the velocity to get a position vector.

$$\mathbf{v} = v_x\mathbf{i} + v_y\mathbf{j} + v_z\mathbf{k}$$

$$\begin{aligned}\mathbf{r} &= \int \mathbf{v}dt \\ &= \int v_x dt\mathbf{i} + \int v_y dt\mathbf{j} + \int v_z dt\mathbf{k}\end{aligned}$$

Note that when integrating like this you will introduce a number of constants of integration. It is important to determine each of these using the initial velocity and initial position or other similar information.

6

Worked example 6.8

A particle moves so that its acceleration, $\mathbf{a}$ m s^{-2}, at time t s is given by:

$$\mathbf{a} = 0.6t\mathbf{i} + (1 - 1.2t)\mathbf{j}$$

where **i** and **j** are perpendicular unit vectors.

(a) Find the velocity of the particle at time t if it has an initial velocity of $(2\mathbf{i} + 3\mathbf{j})$ m s^{-1}.

(b) Find an expression for the position of the particle at time t if its initial position is $20\mathbf{i}$ m.

(c) Find the speed of the particle when $t = 2$.

Solution

(a) To find the velocity at time t integrate the acceleration vector:

$$\begin{aligned}\mathbf{v} &= \int 0.6t dt\mathbf{i} + \int (1 - 1.2t)dt\mathbf{j} \\ &= (0.3t^2 + c_1)\mathbf{i} + (t - 0.6t^2 + c_2)\mathbf{j}\end{aligned}$$

Using the initial velocity, $2\mathbf{i} + 3\mathbf{j}$, gives $c_1 = 2$ and $c_2 = 3$, so that the velocity is:

$$\mathbf{v} = (0.3t^2 + 2)\mathbf{i} + (t - 0.6t^2 + 3)\mathbf{j}$$

(b) To find the position at time t integrate the velocity vector with respect to t:

$$\mathbf{r} = \int (0.3t^2 + 2)dt\,\mathbf{i} + \int (t - 0.6t^2 + 3)dt\mathbf{j}$$

$$= (0.1t^3 + 2t + c_3)\mathbf{i} + (0.5t^2 - 0.2t^3 + 3t + c_4)\mathbf{j}$$

Using the initial position of 20**i** gives $c_3 = 20$ and $c_4 = 0$, so that the position is given by:

$$\mathbf{r} = (0.1t^3 + 2t + 20)\mathbf{i} + (0.5t^2 - 0.2t^3 + 3t)\mathbf{j}$$

(c) When $t = 2$:

$$\mathbf{v} = (0.3 \times 2^2 + 2)\mathbf{i} + (2 - 0.6 \times 2^2 + 3)\mathbf{j}$$

$$= 3.2\mathbf{i} + 2.6\mathbf{j}$$

Then the speed v is given by:

$$v = \sqrt{3.2^2 + 2.6^2}$$

$$= 4.12 \text{ m s}^{-1}$$

EXERCISE 6C

1 A particle moves so that, at time t s, its position vector in metres is given by:

$$\mathbf{r} = (t^2 - 5)\mathbf{i} + (4 - t + 6t^2)\mathbf{j}$$

(a) Find the velocity and acceleration of the particle at time t.

(b) Find the position and velocity of the particle when $t = 4$.

2 Two aeroplanes, A and B, move so that at time t s their position vectors, in metres, are given by

$$\mathbf{r}_A = (30t - 600)\mathbf{i} + (3t^2 - 120t + 1400)\mathbf{j}$$

and

$$\mathbf{r}_B = (20t + 10)\mathbf{i} + (40t - 10)\mathbf{j}$$

where **i** and **j** are unit vectors that are directed east and north, respectively.

(a) Find the velocities of A and B at time t.

(b) Find the speed of B.

(c) Find the time when the two aeroplanes are travelling in parallel directions and the distance between them at this time.

3 A ball rolls on a slope so that its position is given by $\mathbf{r} = (t^2\mathbf{i} + 2t\mathbf{j})$ m at time t s, where **i** and **j** are perpendicular unit vectors. Find the velocity and acceleration the ball at time t.

4 A light aircraft moves so that its position, in metres, relative to an origin O, at time t seconds is given by

$$\mathbf{r} = \left(4t - \frac{t^2}{5}\right)\mathbf{i} + 10t\mathbf{j}$$

where **i** and **j** are unit vectors that are directed east and north, respectively.

(a) Find an expression for the velocity of the aircraft at time t.

(b) Find the time when the aircraft is due north of its initial position, and the distance from its initial position at that time.

(c) Find the time when the aircraft is travelling north and its speed at that time.

(d) Describe fully the acceleration of the aircraft.

5 The position, in metres, of a particle at time t s is given by:

$$\mathbf{r} = (t^2 - 8t + 2)\mathbf{i} + (2t^3 - 5t^2 + 6t)\mathbf{j}$$

where **i** and **j** are horizontal and vertical unit vectors, respectively. The mass of the particle is 3 kg.

(a) Find an expression for the velocity of the particle at time t.

(b) Find an expression for the resultant force acting on the particle at time t.

(c) Find when the horizontal component of the velocity is zero, and the position of the particle at this time.

6 The position, at time t s, of a car overtaking a lorry is modelled, in metres, as $\mathbf{r} = 20t\mathbf{i} + 5\left(t - \frac{t^2}{10}\right)\mathbf{j}$ where **i** and **j** are unit vectors parallel and perpendicular to the straight path of the lorry. The lorry travels along a straight line and has position, in metres, given by $\mathbf{r} = (10 + 15t)\mathbf{i}$ at time t s. Both the car and the lorry are modelled as particles.

(a) Find the time when the car is level with the lorry and its speed at this time.

(b) Find the time when the car is travelling parallel to the lorry and its acceleration at this time.

7 A force of magnitude $5t\mathbf{i} + 10t\mathbf{j}$ N acts on a body, of mass 50 kg, at time t s, for $0 \leqslant t \leqslant 5$. No force acts on the body for $t > 5$. Find the displacement of the body when $t = 10$, if it has an initial velocity of $3\mathbf{i}$ m s^{-1} and starts at the origin. The unit vectors **i** and **j** are perpendicular.

8 A particle moves so that its velocity, in m s^{-1}, at time t s is $\mathbf{v} = (4t^2 + 3)\mathbf{i} + (37.5 - 15t)\mathbf{j}$, where **i** and **j** are perpendicular unit vectors. Initially the particle is at the origin. When the particle is moving parallel to the unit vector **i** the magnitude of the resultant force acting on the particle is 80 N.

(a) Find the mass of the particle.

(b) Find the position of the particle when it is moving parallel to the unit vector **i**.

9 A jet ski has an initial velocity of $(2\mathbf{i} + 5\mathbf{j})$ m s^{-1} and experiences an acceleration of $(\mathbf{i} + 0.2t\mathbf{j})$ m s^{-2}, at time t s. Find expressions for the velocity and position of the jet ski at time t s. Assume that the jet ski starts at the origin.

10 The acceleration of a particle at time t s is $(2t\mathbf{i} - 5t\mathbf{j})$ m s^{-2}. Initially the particle is at the origin and has velocity $(3\mathbf{i} + 6\mathbf{j})$ m s^{-1}.

(a) Find an expression for the velocity of the particle at time t.

(b) Find an expression for the position of the particle at time t.

11 The acceleration of a particle at time t s is $(4\mathbf{i} - t\mathbf{j})$ m s^{-2}. The particle is initially at rest at the point with position vector $(5\mathbf{i} - 10\mathbf{j})$ m. Note that **i** and **j** are perpendicular unit vectors.

(a) Find an expression for the velocity of the particle at time t.

(b) Find an expression for the position of the particle at time t.

12 An object that describes a circular path has a position vector, in metres, at time t s given by $\mathbf{r} = 4\sin(8t)\mathbf{i} + 4\cos(8t)\mathbf{j}$. Find the magnitude of the velocity and acceleration of the object.

13 A particle follows a path so that its position at time t is given by

$$\mathbf{r} = 4\cos(2t)\mathbf{i} + 3\sin(2t)\mathbf{j}.$$

(a) Find the position, velocity and acceleration of the particle when $t = \frac{\pi}{2}$.

(b) Show that the magnitude of the acceleration at time t is

$$\sqrt{(144 + 112\cos^2(2t))}$$

and find its maximum and minimum values. [A]

14 The position vector of smoke particles as they leave a chimney for the first 4 s of their motion is given by

$$\mathbf{r} = 4t\mathbf{i} + \left(\frac{3t^2}{2}\right)\mathbf{j} + 6t\mathbf{k},$$

where **i** and **k** are horizontal, directed north and east, respectively, and **j** is vertically upward.

(a) What is the magnitude and direction of the acceleration of the smoke?

(b) What is
(i) the velocity, and
(ii) the speed
of the smoke particles after 2 s?

(c) Calculate the angle between the direction of the smoke and the ground when $t = 2$.

15 A glider spirals upwards in a thermal (hot air current) so that its position vector with respect to a point on the ground is

$$\mathbf{r} = \left(100\cos\frac{t}{5}\right)\mathbf{i} + \left(200 + \frac{t}{3}\right)\mathbf{j} + \left(100\sin\frac{t}{5}\right)\mathbf{k}.$$

The directions of **i, j** and **k** are as defined in Question **14**.

(a) Determine the glider's speed at $t = 0$, 5π and 10π seconds. What do you notice?

(b) Find **r** when $t = 0$ and 10π seconds and find the height risen in one complete turn of the spiral.

16 A particle moves in a horizontal plane and its position vector at time t is relative to a fixed origin O is given by

$$\mathbf{r} = (2\sin t\mathbf{i} + \cos t\mathbf{j})\text{ m}.$$

Find the values of t in the range $0 \leqslant t \leqslant \pi$ when the speed of the particle is a maximum. [A]

17 The diagram shows the path of a car that overtakes a lorry on a straight road.

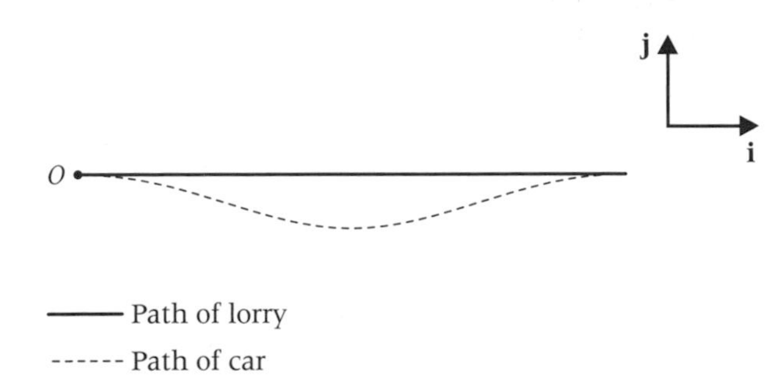

Two alternative models for the position of the car during the 10 s that it takes to pass the lorry are:

Model A $\quad \mathbf{r} = 20t\mathbf{i} - 1.6\left(t - \frac{t^2}{10}\right)\mathbf{j};$

Model B $\quad \mathbf{r} = 20t\mathbf{i} - 2\left(1 - \cos\frac{\pi t}{5}\right)\mathbf{j}.$

(a) For each model find the position and velocity of the car when $t = 0$, 5 and 10.

(b) Which is the better model for the motion of the car? Briefly state your main reason. [A]

18 A possible model for the position of a car, at time t s, while it is travelling over a small hill is:

$$\mathbf{r} = 25t\mathbf{i} + \frac{50}{\pi}\left(1 - \cos\left(\frac{\pi t}{10}\right)\right)\mathbf{j}$$

where **i** and **j** are horizontal and vertical unit vectors, respectively, and the distances are in metres. The model is valid for $0 \leqslant t \leqslant 20$. The diagram shows the path of the car. Initially the car is at O, time $t = 0$.

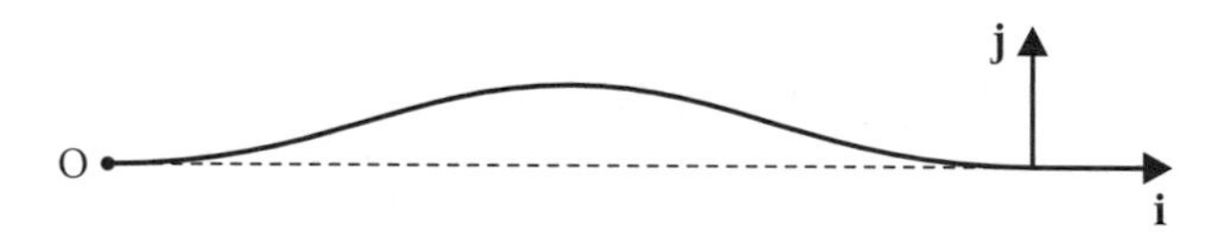

(a) Find the velocity of the car at time t.

(b) Find an expression for the speed of the car, at time t.

(c) Find the times when the car has its minimum and maximum speeds and describe what happens to the speed of the car as it travels over the hill.

(d) Comment on how your answer to (**c**) agrees or disagrees with the way that you would expect the speed of a car to change while going over a hill. [A]

Key point summary

1 In one, two or three dimensions displacements or position vectors can be differentiated to give velocities and accelerations. *p87*

2 Accelerations can be integrated to give velocities and position vectors or displacements. *p87*

Formulae to learn

$$v = \frac{dx}{dt}$$

$$a = \frac{dv}{dt}$$

$$a = \frac{d^2x}{dt^2}$$

$$v = \int a dt$$

$$s = \int v dt$$

$$\mathbf{r} = x\mathbf{i} + y\mathbf{j} + z\mathbf{k}$$

$$\mathbf{v} = \frac{d\mathbf{r}}{dt}$$

$$= \frac{dx}{dt}\mathbf{i} + \frac{dy}{dt}\mathbf{j} + \frac{dz}{dt}\mathbf{k}$$

$$\mathbf{a} = \frac{d\mathbf{v}}{dt}$$

$$= \frac{d^2x}{dt^2}\mathbf{i} + \frac{d^2y}{dt^2}\mathbf{j} + \frac{d^2z}{dt^2}\mathbf{k}$$

$$\mathbf{v} = \int \mathbf{a} dt$$

$$\mathbf{r} = \int \mathbf{v} dt$$

Test yourself	What to review
1 The displacement, s m, of a particle at time t s is given by $s = 5t + 6e^{-2t} - 4$.	*Section 6.2*

1 The displacement, s m, of a particle at time t s is given by $s = 5t + 6e^{-2t} - 4$.

(a) Find the initial velocity of the particle.

(b) Sketch a velocity–time graph for the particle.

(c) Find the acceleration at time t s.

Section 6.2

---

2 A particle moves so that at time t s its acceleration is $-\left(\frac{\sqrt{t}}{4}\right)$ m s^{-2}. Its initial velocity is 16 m s^{-1}, when it passes the origin.
Find the displacement of the particle from the origin when it comes to rest.

Section 6.3

---

3 A particle, of mass 5 kg, moves so that its position vector at time t is given by:

$$\mathbf{r} = (6 + 2t)\mathbf{i} + (8 - 3t^2)\mathbf{j}$$

where $\mathbf{i}$ and $\mathbf{j}$ are perpendicular unit vectors.

(a) Find the position of the particle when $t = 10$.

(b) Find the velocity of the particle in terms of t.

(c) Find the speed of the particle when $t = 2$.

(d) Find the acceleration of the particle in terms of t.

Section 6.4

---

4 A particle moves so that its acceleration, $\mathbf{a}$ m s^{-2}, at time t s is given by

$$\mathbf{a} = 0.6t\mathbf{i} + (1 - 1.2t)\mathbf{j}$$

where $\mathbf{i}$ and $\mathbf{j}$ are perpendicular unit vectors.

(a) Find the velocity of the particle at time t if it has an initial velocity of $(2\mathbf{i} + 3\mathbf{j})$ m s^{-1}.

(b) Find an expression for the position of the particle at time t if its initial position is $20\mathbf{i}$.

(c) Find the speed of the particle when $t = 2$.

Section 6.4

Test yourself ANSWERS

1 (a) -7 m s^{-1}, **(c)** $a = 24e^{-2t}$.

2 201 m.

3 (a) $26\mathbf{i} - 292\mathbf{j}$, **(b)** $2\mathbf{i} - 6t\mathbf{j}$, **(c)** 12.2, **(d)** $-6\mathbf{j}$.

4 (a) $\mathbf{v} = (2 + 0.3t^2)\mathbf{i} + (3 + t - 0.6t^2)\mathbf{j}$.

(b) $\mathbf{r} = (20 + 2t + 0.1t^3)\mathbf{i} + (3t + 0.5t^2 - 0.2t^3)\mathbf{j}$.

(c) 4.12 m s^{-1}.

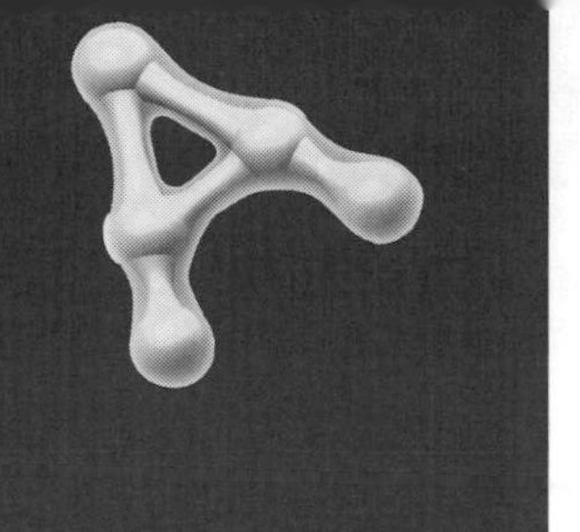

CHAPTER 7

Energy

Learning objectives

After studying this chapter you should be able to:

- calculate kinetic energy
- calculate work done by a constant force
- calculate gravitational potential energy
- calculate elastic potential energy
- be able to use power.

7.1 Kinetic energy

Every moving body has kinetic energy. The greater the mass and the greater the speed, the greater the kinetic energy.

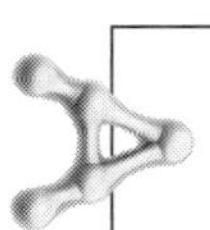

> The kinetic energy of a body is defined as $\frac{1}{2}mv^2$, where m is the mass of the body and v its speed.

The units of energy are joules (J).

Worked example 7.1

A car has mass 1100 kg. At the bottom of a hill it is travelling at 30 m s^{-1} and loses speed as it travels up the hill. At the top of the hill its speed is 22 m s^{-1}. Calculate the amount of kinetic energy lost as the car drove up the hill.

Solution

At the bottom of the hill;

$$\text{Kinetic energy} = \tfrac{1}{2} \times 1100 \times 30^2$$
$$= 495\,000 \text{ J}$$

At the top of the hill;

$$\text{Kinetic energy} = \tfrac{1}{2} \times 1100 \times 22^2$$
$$= 266\,200 \text{ J}$$

Now the amount of kinetic energy that has been lost can be calculated.

$$\text{Kinetic energy lost} = 495\,000 - 266\,200$$
$$= 228\,800 \text{ J}$$

EXERCISE 7A

1 Calculate the kinetic energy of a ball, of mass 150 g, travelling at 8 m s^{-1}.

2 Calculate the kinetic energy of a train, of mass 30 000 tonnes, travelling at 50 m s^{-1}.

3 A ball has a mass of 200 g. It is thrown so that its initial speed is 12 m s^{-1} and during its flight it has a minimum speed of 6 m s^{-1}. Calculate the minimum and maximum values of the kinetic energy of the ball.

4 A light aeroplane has a mass of 1500 kg. When it lands it is travelling at 80 m s^{-1} and at the end of the runway its speed has been reduced to 10 m s^{-1}. Calculate how much kinetic energy has been lost.

5 A stone, of mass 50 g, is dropped from the top of a cliff at a height of 40 m.

(a) Assume that no resistance forces act on the stone and calculate its speed at the bottom of the cliff.

(b) How much kinetic energy does the stone gain as it falls.

6 A cycle and cyclist have mass 70 kg. The cyclist freewheels from rest down a slope, accelerating at 0.5 m s^{-2}. The initial speed of the cyclist is 3 m s^{-1}.

(a) Calculate the speed of the cyclist after he has travelled 50 m.

(b) Calculate the increase in the kinetic energy of the cyclist.

7.2 Work and energy

As a stone falls its kinetic energy increases. As you start to pedal a cycle your kinetic energy increases. In the first of these cases gravity is the force that causes a change in kinetic energy. In the second the cyclist exerts a force. In this section you will examine the relationship between the change in the kinetic energy of a body and the forces that act on it.

If a constant force of magnitude F acts on a body of mass m, it will produce an acceleration of $\frac{F}{m}$, this can be substituted into the constant acceleration equation $v^2 = u^2 + 2as$, to give

$$v^2 = u^2 + 2 \times \frac{F}{m} \times s$$

or

$$\tfrac{1}{2}mv^2 - \tfrac{1}{2}mu^2 = Fs$$

This equation can be expressed as

$$\text{Change in kinetic energy} = Fs$$

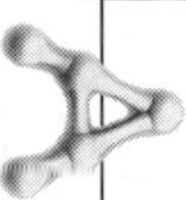

The quantity Fs is referred to as the work done by the force. It is this work that determines the change in the kinetic energy of the body that the force causes.

Worked example 7.2

A ball, of mass 0.4 kg, is released from rest and allowed to fall 3 m.

(a) Find the work done by gravity as the ball falls.

(b) State the gain in kinetic energy of the ball.

(c) Calculate the speed of the ball when it has fallen 3 m.

Solution

(a) The work done is calculated using Fs. In this case

$$\begin{aligned} F &= mg \\ &= 0.4 \times 9.8 \\ &= 3.92 \text{ N} \end{aligned}$$

The work done by gravity can be calculated.

$$\begin{aligned} \text{Work done} &= Fs \\ &= 3.92 \times 3 \\ &= 11.76 \text{ J} \end{aligned}$$

(b) As the gain in kinetic energy is equal to the work done, there is a gain in kinetic energy of 11.76 J.

(c) The kinetic energy of the ball is 11.76 J, so the speed can be calculated as below:

$$\begin{aligned} 11.76 &= \tfrac{1}{2} \times 0.4 v^2 \\ v^2 &= 58.8 \\ v &= \sqrt{58.8} \\ &= 7.67 \text{ m s}^{-1} \text{ (to 3 sf)} \end{aligned}$$

Worked example 7.3

A box is initially at rest on a smooth horizontal surface. The mass of the box is 5 kg. A horizontal force of magnitude 8 N acts on the box as it slides 6 m.

(a) Find the work done by the force.

(b) Find the speed of the box when it has travelled 6 m.

Solution

(a) The work done is calculated using Fs.

$$\text{Work done} = 8 \times 6 = 48 \text{ J}$$

(b) Using the fact that the work done is equal to the change in kinetic energy gives

$$48 = \tfrac{1}{2} \times 5v^2$$
$$v = \sqrt{19.2}$$
$$= 4.38 \text{ m s}^{-1} \text{ (to 3 sf)}$$

In most situations more than one force will act. Some forces may act in the direction of motion, as in the previous examples, but often they will act in the opposite direction to the motion. These types of forces will include resistance and friction forces. A force that acts in the opposite direction to the motion will do a negative amount of work. For example the work done by a friction force of magnitude 80 N acting on a body that moves 5 m would be −400 J. Often we would say that the work done against friction is 400 J.

Worked example 7.4

A car, of mass 1250 kg, is subject to a forward force of magnitude 2000 N and a resistance force of magnitude 500 N. The car moves 200 m.

(a) Find the work done by each of the forces acting on the car.

(b) If the car is initially moving at 5 m s^{-1}, find the final speed of the car.

Solution

(a) The work done by the 2000 N force is $2000 \times 200 = 400\,000$ J.

The work done by the 500 N force is $-500 \times 200 = -100\,000$ J. We might say that the work done against friction is 100 000 J. Total work done = 400 000 − 100 000 = 300 000 J

Alternatively note that the resultant is 1500 N and the work done will be $1500 \times 200 = 300\,000$ J.

(b) The change in kinetic energy is 300 000 J. So the final speed can be calculated

$$300\,000 = \tfrac{1}{2} \times 1250 \times v^2 - \tfrac{1}{2} \times 1250 \times 5^2$$
$$v^2 = 505$$
$$v = \sqrt{505}$$
$$= 22.5 \text{ m s}^{-1} \text{ (to 3 sf)}$$

Worked example 7.5

A ball, of mass 0.3 kg, is moving at 8 m s^{-1} when it enters a tank of water. It hits the bottom of the tank travelling at 2 m s^{-1}. The depth of water in the tank is 1.2 m. Assume that a constant resistance force acts on the ball as it moves through the water.

(a) Calculate the change in the kinetic energy of the ball.

(b) Find the magnitude of the resistance force that acts on the ball.

Solution

(a) Change in kinetic energy $= \frac{1}{2} \times 0.3 \times 2^2 - \frac{1}{2} \times 0.3 \times 8^2$
$= -9 \text{ J}$

(b) First consider the work done by each of the forces.

$$\text{Work done by gravity} = 0.3 \times 9.8 \times 1.2 = 3.528 \text{ J}$$

If the resistance force has magnitude R, then the work done by this force will be

$$-R \times 1.2 = -1.2R$$

and so the total work done is

$$3.528 - 1.2R.$$

Now we can find R by using

$$\text{Change in kinetic energy} = \text{work done}$$

$$-9 = 3.528 - 1.2R$$

$$R = \frac{3.528 + 9}{1.2}$$

$$= 10.44 \text{ N}$$

EXERCISE 7B

1 A force of magnitude 800 N acts on a car, of mass 1000 kg, as it moves 400 m on a horizontal surface. Assume that no other forces act on the car.

(a) Calculate the work done by the force that acts on the car.

(b) Find the final kinetic energy and speed of the car if it is initially

(i) at rest,

(ii) moving at 3 m s^{-1}.

2 A force acts horizontally on a package, of mass 4 kg, that is initially at rest on a smooth horizontal surface. After the package has moved 3 m its speed is 5 m s^{-1}.

(a) Find the increase in the kinetic energy of the package.

(b) How much work is done by the force that acts on the package?

(c) Determine the magnitude of the force that acts on the package.

3 A brick, of mass 2 kg, is allowed to fall from rest at a height of 3.2 m. Find the kinetic energy and speed of the brick when it hits the ground:

(a) assuming that no resistance forces act on the brick as it falls,

(b) assuming that a constant resistance force of magnitude 10 N acts on the brick.

4 A rope is attached to a boat, of mass 200 kg. The boat is pulled along a horizontal surface by the horizontal rope. The tension in the rope remains constant at 500 N. When the boat has moved 50 m its speed is 0.4 m s^{-1}.

(a) Calculate the work done by the tension in the rope.

(b) Calculate the final kinetic energy of the boat.

(c) Find the work done against the resistance forces acting on the boat.

(d) Find the magnitude of the resistance force if it is assumed to be constant.

5 A ball of mass 200 g is dropped from a height of 1 m. The ball is initially at rest. It hits the ground and rebounds at $\frac{3}{4}$ of the speed with which it hit the ground. Find the energy lost by the ball as it bounces and the height to which it rebounds.

6 A forward force of magnitude 2500 N acts on a car, of mass 1250 kg, as it moves 500 m. The initial speed of the car was 12 m s^{-1}. All the motion takes place along a straight line.

(a) If no resistance forces act on the car find its final kinetic energy and speed.

(b) If the final speed of the car is 30 m s^{-1}, find the work done against the resistance forces and the average magnitude of the resistance force.

7 As a car, of mass 1200 kg, skids 25 m, on a horizontal surface, its speed is reduced from 30 m s^{-1} to 20 m s^{-1}.

(a) Find the energy lost by the car as it skids.

(b) If the coefficient of friction between the car and the road is 0.8, find the work done by the friction force.

(c) Find the work done by the air resistance force that acts on the car.

(d) Assuming that the air resistance force is constant, find out how much further the car travels before it stops.

(e) Criticise the assumptions that you have used to get your answer to **(d)** and suggest how using a more realistic model would change your answer.

8 A diver, of mass 60 kg, dives from a diving board at a height of 4 m. She hits the water travelling at a speed of 8 m s^{-1} and descends to a depth of 2 m in the diving pool. Model the diver as a particle.

(a) Find the work done against air resistance before the diver hits the water and the average magnitude of the air resistance force.

(b) Find the average magnitude of the force exerted by the water on the diver as she is brought to rest in the diving pool.

9 A ball of mass 500 g is dropped from a height of 3 m. It hits the ground and travels down 6 cm into soft mud before stopping. Assume no air resistance acts on the ball as it falls.

(a) Find the average force that the mud exerts on the ball while it is slowing down;

(b) A different ball of mass 200 g hits the mud travelling at 6 m s^{-1}. Assume that the mud exerts the same force on the ball and find the distance that it travels into the mud before stopping.

10 A bullet fired into an earth bank at an unknown speed penetrates to a distance of 2 m. An identical bullet fired at a speed of 120 m s^{-1} travels 3 m into the bank. If the mass of this type of bullet is 20 g, find the average force exerted on the bullets by the earth, and the speed at which the first bullet was fired.

7.3 Forces at angles

If a force acts at an angle to the direction of motion to the body that it acts on, then we must use the component of the force in the direction of motion when calculating the work done.

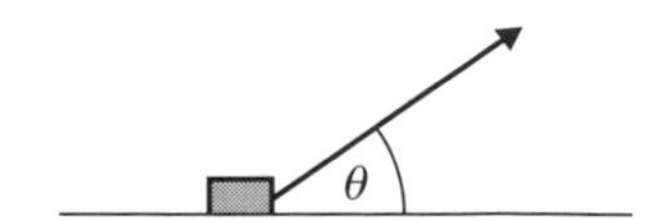

If the force shown in the diagram acts as the body moves a distance d, then the work done is $Fd \cos \theta$.

Worked example 7.6

A force, of magnitude 60 N, acts at an angle of 30° above the horizontal on a sack, of mass 50 kg, that is initially at rest on a horizontal surface. Find the work done by the force and the final speed of the sack which moves 5 m when:

(a) the surface is smooth,

(b) the coefficient of friction between the sack and the surface is 0.1.

Solution

(a) The work done can be calculated using $Fd\cos\theta$.

$$\begin{aligned}\text{Work done} &= 60 \times 5 \times \cos 30° \\ &= 300 \times \frac{\sqrt{3}}{2} \\ &= 150\sqrt{3}\text{ J}\end{aligned}$$

The work done is equal to the gain in kinetic energy, so

$$\begin{aligned}150\sqrt{3}\text{ J} &= \tfrac{1}{2} \times 50v^2 \\ v^2 &= 6\sqrt{3} \\ v &= 3.22\text{ m s}^{-1}\text{ (to 3 sf)}\end{aligned}$$

(b) The work done by the force will be the same, but the final kinetic energy will be less due to the friction force. First we must calculate the magnitude of the friction force.

Resolving vertically to find the normal reaction, R, we have

$$\begin{aligned}R + 60\sin 30° &= 50 \times 9.8 \\ R &= 460\text{ N}\end{aligned}$$

Then the magnitude of the friction force can now be found using $F = \mu R$, which in this case gives

$$\begin{aligned}F &= 0.1 \times 460 \\ &= 46\text{ N}\end{aligned}$$

The work done by the friction force will be

$$-46 \times 5 = -230\text{ J}$$

The total work done is then $150\sqrt{3} - 230$.
The speed can now be found using work done equals change in kinetic energy.

$$\begin{aligned}150\sqrt{3} - 230 &= \tfrac{1}{2} \times 50v^2 \\ v &= 1.09\text{ m s}^{-1}\text{ (to 3 sf)}\end{aligned}$$

Worked example 7.7

A particle, of mass 5 kg, is initially at rest. It slides 4 m down a slope at 30°. Assume that the slope is smooth and that there is no air resistance.

(a) Find the work done by gravity as the particle slides down the slope.

(b) Find the speed of the particle, when it has travelled the 4 m.

Solution

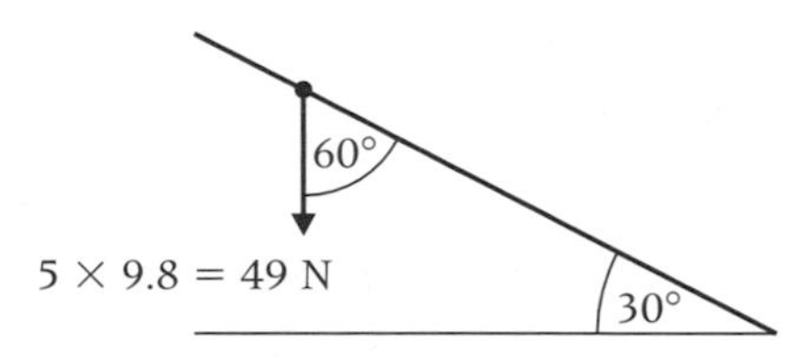

(a) The work done can be calculated using $Fd \cos \theta$.

$$\text{Work done} = 5 \times 9.8 \times 4 \cos 60° = 98 \text{ J}$$

(b) Using work done equals change in kinetic energy gives

$$98 = \tfrac{1}{2} \times 5v^2$$
$$v = \sqrt{39.2} = 6.26 \text{ m s}^{-1} \text{ (to 3 sf)}$$

Gravitational potential energy

Consider a particle of mass m that falls vertically a distance h, in the absence of any resistance forces. The work done by gravity will be given by mgh.

Also consider a second particle of the same mass that slides down the smooth slope shown in the diagram.

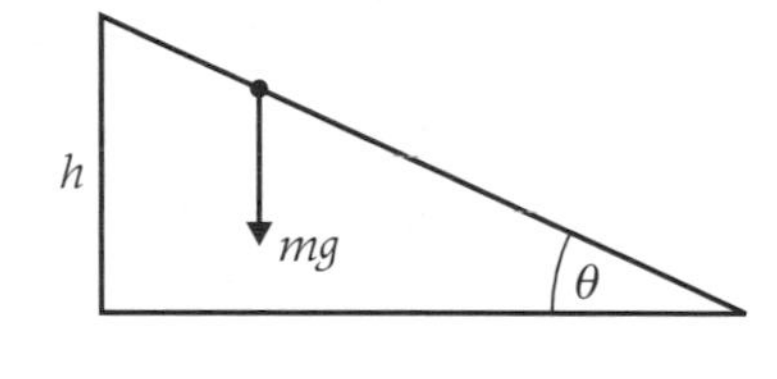

The particle will slide a distance of $\dfrac{h}{\sin \theta}$.

So the work done by gravity is

$$mg \times \frac{h}{\sin \theta} \times \cos (90 - \theta) = mg \times \frac{h}{\sin \theta} \times \sin \theta = mgh$$

Note that in both of these examples the work done by gravity is the same and depends on the initial height and not the route taken. A curved surface would have also produced the same result. The important feature of this result is that there are no resistance or friction forces present.

This quantity mgh is often referred to as the gravitational potential energy of the body and often this is abbreviated to potential energy or PE. If a body is allowed to fall, swing or slide the potential energy will be converted to kinetic energy. Similarly as a body rises its kinetic energy will be converted to potential energy. This gives a useful way of approaching some problems, because the total energy will remain constant if no resistance or friction forces act.

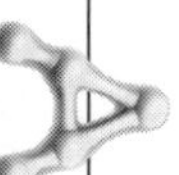

Gravitational potential energy $= mgh$

Worked example 7.8

A soldier, of mass 72 kg, on a training exercise is running at a speed of 5 m s^{-1}. He grabs hold of a rope, of length 6 m, that is hanging vertically. He then swings on the rope.

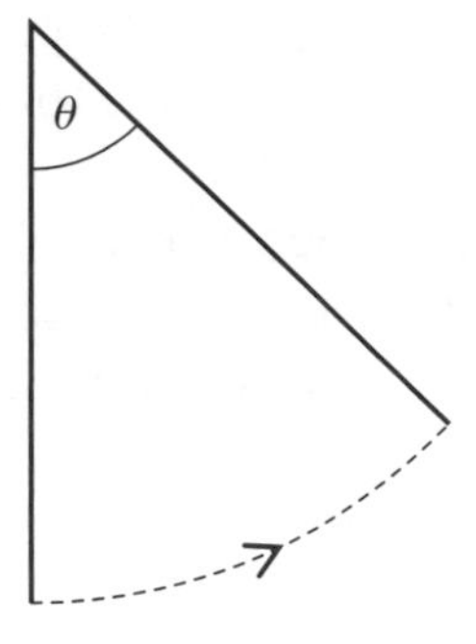

(a) Calculate the initial kinetic energy of the soldier.

(b) Calculate the speed of the soldier when he has risen 0.5 m.

(c) Find the maximum height of the soldier and the angle between the rope and the vertical at this time.

Solution

(a) The initial kinetic energy can be calculated from the information given

$$\text{KE} = \tfrac{1}{2} \times 72 \times 5^2$$
$$= 900 \text{ J}$$

(b) When the soldier has risen 0.5 m his potential energy can be calculated

$$\text{PE} = 72 \times 9.8 \times 0.5$$
$$= 352.8 \text{ J}$$

The remaining kinetic energy can then be calculated as

$$900 - 352.8 = 547.2$$

The speed of the soldier can now be found

$$547.2 = \tfrac{1}{2} \times 72v^2$$
$$v = \sqrt{15.2}$$
$$= 3.90 \text{ m s}^{-1} \text{ (to 3 sf)}$$

(c) At the soldier's highest point, all his initial kinetic energy will have been converted to potential energy. This gives the equation

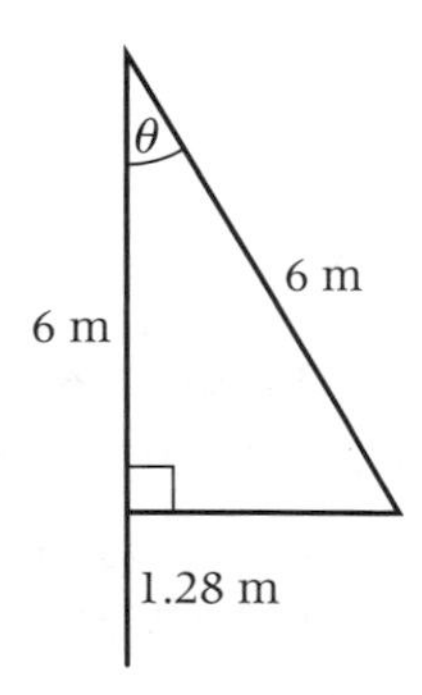

$$72 \times 9.8h = 900$$
$$h = 1.28 \text{ m (to 3 sf)}$$

By considering the triangle in the diagram

$$\cos\theta = \frac{6 - 1.28}{6}$$

$\theta = 38°$ to the nearest degree.

Worked example 7.9

A cyclist and cycle of combined mass 80 kg freewheel down a slope. They travel a distance of 100 m down the slope which is at an angle α to the horizontal, where $\sin\alpha = \frac{1}{10}$. The speed of the cyclist increases from 4 m s^{-1} to 8 m s^{-1}.

(a) Find the change in the total energy of the cycle and cyclist at the top and bottom of the slope.

(b) Find the work done against resistance forces while the cyclist travelled down the slope and the magnitude of the average resistance force on the cyclist.

(c) Find the speed of the cyclist when he has travelled 20 m down the slope.

Solution

(a) We must consider the gain in kinetic energy and the potential energy that is lost.

$$\begin{aligned}\text{Gain in KE} &= \tfrac{1}{2} \times 80 \times 8^2 - \tfrac{1}{2} \times 80 \times 4^2 \\ &= 1920\text{ J}\end{aligned}$$

$$\begin{aligned}\text{Loss of PE} &= 80 \times 9.8 \times 100 \sin \alpha \\ &= 80 \times 9.8 \times 100 \times \tfrac{1}{10} \\ &= 7840\text{ J}\end{aligned}$$

$$\begin{aligned}\text{Change in energy} &= 1920 - 7840 \\ &= -5920\text{ J}\end{aligned}$$

So 5920 J of energy has been lost.

(b) The energy has been lost due to the work done against friction, so

$$\text{Work done against resistance forces} = 5920\text{ J}$$

The average resistance force can be found by dividing the work done by the distance travelled.

$$\text{Average resistance force} = \frac{5920}{100} = 59.2\text{ N}$$

(c) As the cycle and cyclist travel the 20 m, the resistance force will act. So first calculate the work done against these forces.

$$\text{Work done against resistance forces} = 20 \times 59.2 = 1184\text{ J}$$

As the cycle and cyclist travel down the slope more potential energy is lost.

$$\text{Loss of PE} = 80 \times 9.8 \times 20 \times \tfrac{1}{10} = 1568\text{ J}$$

The gain in kinetic energy will then be $1568 - 1184 = 384$ J. The initial KE is given by

$$\tfrac{1}{2} \times 80 \times 4^2 = 640\text{ J}$$

The final kinetic energy will then be $640 + 384 = 1024$ J. Now the final speed can be found

$$\begin{aligned}\tfrac{1}{2} \times 80v^2 &= 1024\text{ J} \\ v &= \sqrt{25.6} \\ &= 5.06\text{ m s}^{-1} \text{ (to 3 sf)}\end{aligned}$$

7

EXERCISE 7C

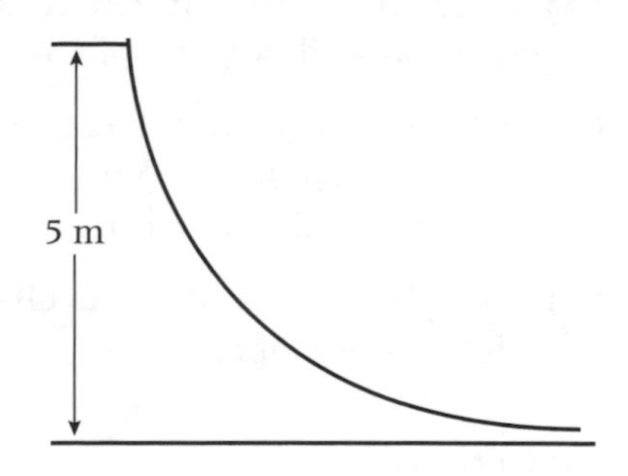

1 The diagram shows a curved slide with a drop of 5 m. A child, of mass 50 kg, sits at the top of the slide. He slides down. Assume that there are no resistance or friction forces acting on the slide. Calculate:

(a) the potential energy that the child would lose as he slides from the top to the bottom of the slide,

(b) the speed of the child at the bottom of the slide.

2 A stone, of mass 0.8 kg, is thrown over a cliff at speed of 3 m s^{-1}. It hits the water at a speed of 12 m s^{-1}.

(a) Find the potential energy lost by the stone.

(b) Find the height of the cliff.

3 A particle, of mass 3 kg, slides down a slope of length 20 m, which is inclined at an angle of 45° to the horizontal. At the top of the slope the particle has an initial speed of 4 m s^{-1}. Assume that the slope is smooth.

(a) Find the potential energy lost by the particle as slides down the slope.

(b) Find the kinetic energy and the speed of the particle at the bottom of the slope.

(c) If a constant friction force of magnitude 5 N acts on the particle as it slides, find the speed of the particle at the bottom of the slope.

4 A child, of mass 60 kg, swings on a rope of length 8 m. The rope is initially at an angle of 30° to the vertical. The child initially moves at 2 m s^{-1}.

(a) Find the potential energy that is lost as the child swings to her lowest point.

(b) Find the maximum kinetic energy and the maximum speed of the child.

(c) Find the maximum height of the child above her lowest position.

5 A ball of mass 300 g is kicked so that it has an initial speed of 12 m s^{-1}. During its flight the speed of the ball has a minimum value of 4 m s^{-1}.

(a) Find the initial kinetic energy of the ball.

(b) Find the maximum potential energy of the ball.

(c) Find the maximum height of the ball.

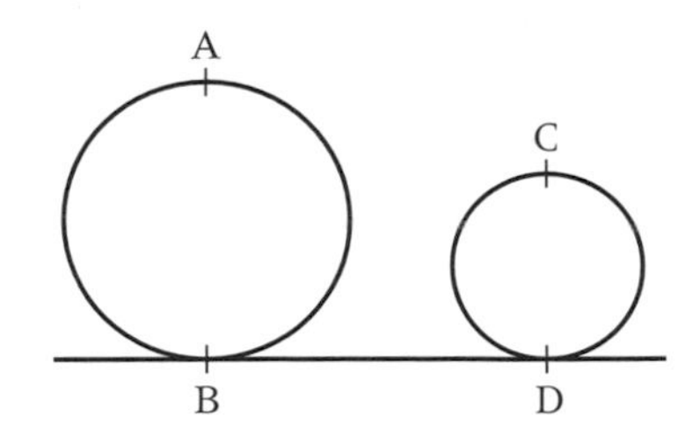

6 A loop-the-loop roller coaster is shown in the diagram. At point A at the top of the first loop the roller coaster is moving at $8\,\mathrm{m\,s^{-1}}$. The mass of the roller coaster carriage is 400 kg. 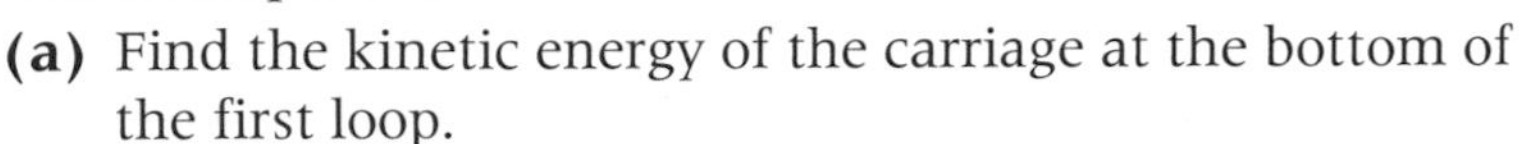 Assume that no resistance forces act on the roller coaster. The diameter of the first loop is 5 m and the diameter of the second loop is 3 m.

(a) Find the kinetic energy of the carriage at the bottom of the first loop.

(b) Find the kinetic energy of the carriage at the top of the second loop.

(c) Find the maximum speed of the roller coaster.

7 A sledge, of mass 12 kg, is pulled by a rope that is at an angle of 20° to the horizontal. The tension in the rope is a constant 80 N. The coefficient of friction between the sledge and the horizontal ground on which it moves is 0.2. Find the kinetic energy and the speed of the sledge when it has moved 5 m from rest.

8 A roller coaster, of mass 500 kg, is at the top of a slope and travelling at $4\,\mathrm{m\,s^{-1}}$. As it travels down the slope its speed increases to $10\,\mathrm{m\,s^{-1}}$. The length of the slope is 20 m and the top is 12 m higher than the bottom. At the bottom of the slope it travels on a horizontal section of track. Model the roller coaster as a particle that has a constant resistance force acting on it.

(a) Find the energy lost by the roller coaster as it moves down the slope.

(b) Find the magnitude of the resistance force on the roller coaster.

(c) Find how far the roller coaster travels along the horizontal section of the track before it comes to rest.

9 A car, of mass 1100 kg, is travelling down a hill, inclined at an angle of 5° to the horizontal. The driver brakes hard and skids 15 m. The coefficient of friction between the tyres and the road is 0.7. Find the initial kinetic energy and speed of the car.

10 The diagram shows a slide that is in the shape of a semicircle of radius 5 m and that has centre O. The users slide down the inside of the slide and up the other side. They all start at rest at the top of the slide, at the point A. The mass of a user is 55 kg.

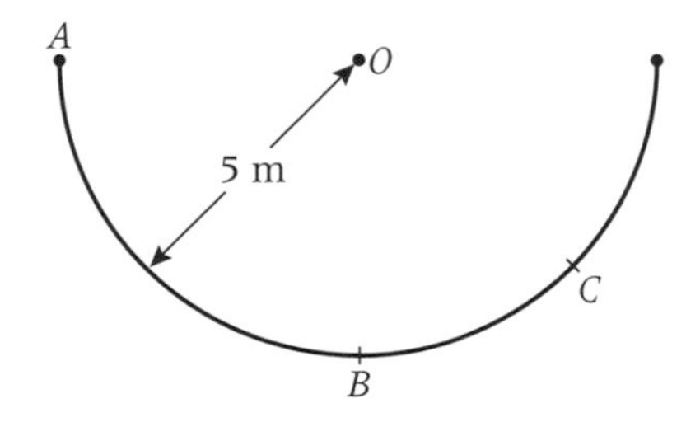

(a) If the slide is smooth find the maximum kinetic energy of a user and their maximum speed.

In fact the slide is not smooth. A simple model assumes that a constant resistance force acts on the users and the magnitude of this is 30 N.

(b) Calculate the speed of a user at the lowest point of the slide, marked B on the diagram.

(c) A person using the slide comes to rest at the point marked C. Find the angle BOC. (Hint: use a numerical method to solve your equation.)

11 A soldier, of mass 80 kg, swings on a rope of length 6 m. He is to be modelled as a particle that describes a circular arc from A, through B to C. The path is shown in the following diagram.

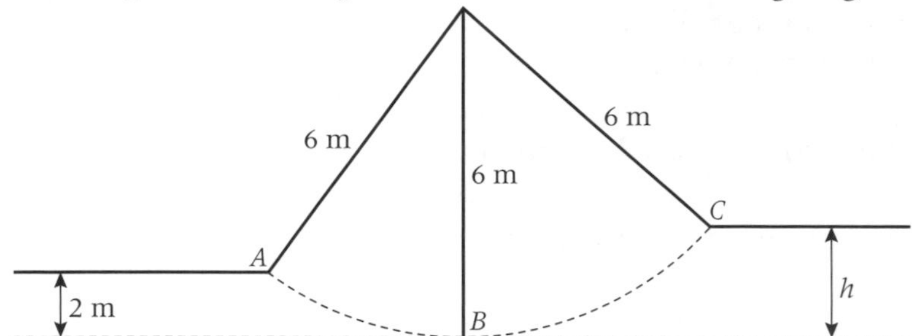

The point A is 2 m higher than B and C is h m higher than B. Initially the soldier moves at $2\ \text{m s}^{-1}$ at A and in a direction perpendicular to the rope.

(a) Find the kinetic energy of the soldier at B, stating any assumptions that you make.

(b) Find h, if the soldier comes to rest at C before swinging back.

(c) Explain why the tension does no work in this situation. [A]

12 The diagram shows part of the track of a roller coaster ride, which has been modelled as a number of straight lengths of track. The roller coaster's carriages are modelled as a particle of mass 400 kg, which can negotiate the bends A, B, C and D without any loss of speed. The speed of the roller coaster at A is $3\ \text{m s}^{-1}$ and at B it is $10\ \text{m s}^{-1}$.

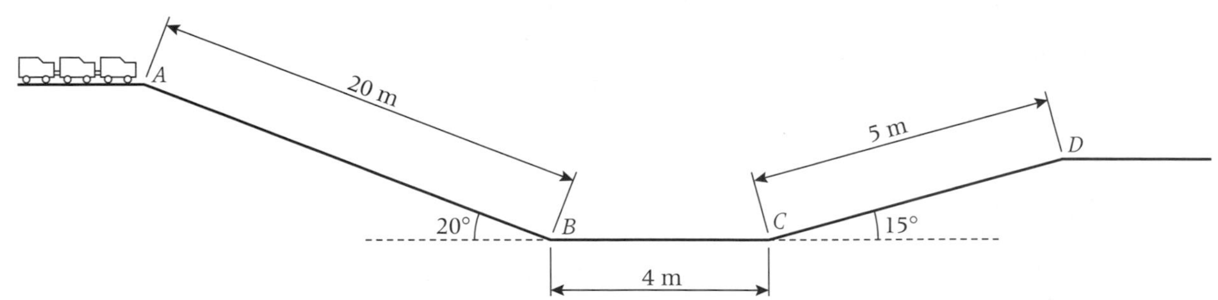

(a) Show that the work done against the resistance forces, as the roller coaster moves from A to B, is approximately 8610 J and use this to find the magnitude of the resistance forces, assuming that they are constant.

(b) Using the magnitude of the resistance forces found in **(a)**, show that the speed of the roller coaster at D is approximately $7.4\ \text{m s}^{-1}$ and find how far along the horizontal track beyond D it could travel before stopping.

(c) Describe **two** ways in which this model could be improved. [A]

7.4 Hooke's law

This section will consider the work done by variable forces, but will first introduce Hooke's law, which predicts the tension in a string or spring. Hooke's law will be used extensively in the later sections of this chapter.

Hooke's law provides a simple but effective model for the tension in a spring.

Hooke's law simply states that the tension T is given by

$$T = \frac{\lambda x}{l}$$

where λ is a constant called the modulus of elasticity, l is the natural or unstretched length of the spring and x is the extension of the spring. The modulus of elasticity depends on the material that the spring is made from and the way in which it has been constructed.

Note that when working with Hooke's law all lengths should be in metres and that the units of λ are newtons.

If a spring is compressed instead of stretched, then Hooke's law can be used to calculate the thrust exerted by the spring.

Hooke's law can also be applied to elastic strings.

Worked example 7.10

A particle, of mass 5 kg, is suspended from a spring, of natural length 0.2 m and modulus of elasticity 40 N. Find the extension of the spring when the particle is in equilibrium.

Solution

The diagram shows the forces acting on the particle. In equilibrium, the upward tension will balance the weight of the particle. This gives

T

$5 \times 9.8 = 49$ N

$$T = 5 \times 9.8$$
$$= 49 \text{ N}$$

Using Hooke's law this becomes;

$$\frac{40x}{0.2} = 49$$
$$x = 0.245 \text{ m}$$

Worked example 7.11

A spring has natural length 0.5 m. A 2 kg mass is suspended from the spring and in equilibrium the extension of the spring is 0.05 m. Find the modulus of elasticity of the spring.

Solution

The diagram shows the forces acting on the mass. When it is in equilibrium, we have;

T

$2 \times 9.8 = 19.6$ N

$$T = 2 \times 9.8$$
$$= 19.6 \text{ N}$$

Using Hooke's law this becomes;

$$\frac{\lambda \times 0.05}{0.5} = 19.6$$
$$\lambda = 196 \text{ N}$$

EXERCISE 7D

1 A spring has modulus of elasticity 40 N and natural length 0.8 m. A particle is attached to the end of the spring and the system is allowed to hang vertically. Find the extension of the spring when the particle is in equilibrium, if the mass of the particle is:

(a) 2 kg,

(b) 1.2 kg,

(c) 200 g.

2 A spring has natural length 0.25 m and modulus of elasticity 20 N. A force of magnitude 40 N is applied to one end, while the other remains fixed. Find the extension of the spring when the forces are in equilibrium.

3 A spring has natural length 20 cm. When it supports a particle of mass 4 kg in equilibrium, it has an extension of 5 cm. Find the modulus of elasticity of the spring.

4 An elastic string has natural length 60 cm and modulus of elasticity 4 N. It stretches 10 cm when it supports an object with an unknown mass in equilibrium. Find the mass of the object.

5 Two identical springs have natural length 8 cm and modulus of elasticity 20 N. A 100 g mass is attached to the springs so that it is in equilibrium.

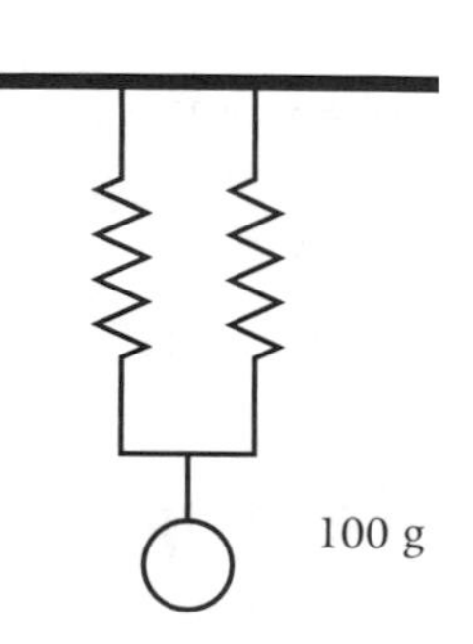

(a) Find the extension of a single spring that supports the mass.

(b) If the springs support the mass as shown in the diagram, find the extension of each spring.

(c) If the springs are joined end to end and then support the mass, find the total extension of springs.

7.5 Energy and variable forces

So far our considerations have been restricted to forces that either remain constant or that are modelled as being constant. In this section you will consider how to extend the ideas previously encountered to variable forces.

If a force that has magnitude $f(x)$, where x is the displacement of the particle, acts on a particle, of mass m, then applying Newton's second law gives

$$f(x) = ma$$

In the chapter on kinetics and variable acceleration you saw that the acceleration can be expressed as $\frac{dv}{dt}$, but using the chain rule this can also be expressed as $\frac{dv}{dx} \times \frac{dx}{dt}$ or $v\frac{dv}{dx}$ and so the equation above could be written as

$$f(x) = mv\frac{dv}{dx}$$

Noting that as x changes from x_1 to x_2, the speed will change from u to v you can integrate to obtain

$$\int_{x_2}^{x_1} f(x)dx = m\int_u^v vdv$$

$$= m\left[\frac{1}{2}v^2\right]_u^v$$

$$= \frac{1}{2}mv^2 - \frac{1}{2}mu^2$$

This result shows that integrating $f(x)$ with respect to x will give the change in kinetic energy and the work done by a variable force is $\int f(x)dx$.

Elastic potential energy

When a spring is stretched, work is done. In the same way as when work is done lifting a body it gains potential energy, a stretched or compressed spring also has potential energy, that could be converted to kinetic energy if the spring is released.

Consider a spring. The tension in the spring is given by $T = \frac{\lambda x}{l}$.

If the spring is extended by a distance e from its natural length, then the work done will be given by

$$\int_0^e \frac{\lambda x}{l}dx = \left[\frac{\lambda x^2}{2l}\right]_0^e$$

$$= \frac{\lambda e^2}{2l}$$

As this is the work done in stretching the spring, you can also state that this is also the amount of potential energy stored in the spring. This is often expressed as follows.

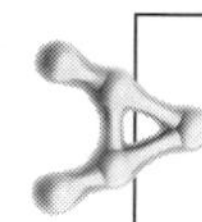

> The elastic potential energy (EPE) of a stretched (or compressed) spring $= \dfrac{\lambda e^2}{2l}$.

The following examples show how this result can be applied.

Worked example 7.12

A spring has natural length 20 cm and modulus of elasticity 80 N. Calculate the work done in stretching the spring

(a) from its natural length to a length of 25 cm,

(b) from a length of 30 cm to 40 cm.

Solution

(a) This is found by substituting $\lambda = 80$, $l = 0.2$ and $e = 0.05$ into the formula $\dfrac{\lambda e^2}{2l}$

$$\begin{aligned}\text{Work done} &= \frac{80 \times 0.05^2}{2 \times 0.2}\\ &= 0.5\ \text{J}\end{aligned}$$

(b) The required amount is the work done to stretch the spring to 40 cm less the work done to stretch it to 30 cm. Note that the extensions will be 0.1 m and 0.2 m.

$$\begin{aligned}\text{Work done} &= \frac{80 \times 0.2^2}{2 \times 0.2} - \frac{80 \times 0.1^2}{2 \times 0.2}\\ &= 8 - 2\\ &= 6\ \text{J}\end{aligned}$$

Worked example 7.13

A ball, of mass 300 grams, is placed on top of a spring of natural length 10 cm and modulus of elasticity 80 N. The spring is compressed until its length is 5 cm and released. Find the maximum height of the ball above the base of the spring.

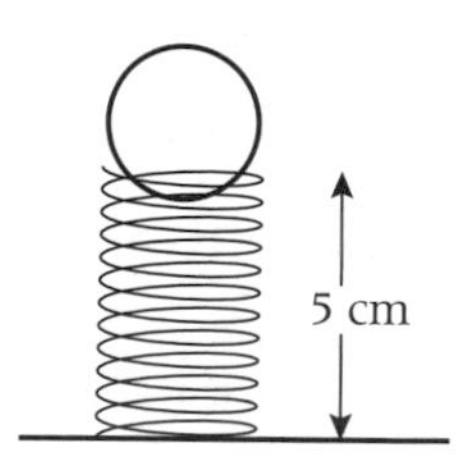

Solution

The initial energies can be calculated.

The EPE of the spring $\dfrac{80 \times 0.05^2}{2 \times 0.1} = 1\ \text{J}$

The gravitational potential energy of the ball is $0.3 \times 9.8 \times 0.05 = 0.147$ J.

So the total initial energy of the ball is 1.147 J.

At its highest point the gravitational potential energy is $0.3 \times 9.8 \times h = 2.94h$ J.

If you assume that energy is conserved, then

$$\begin{aligned}2.94h &= 1.147\\ h &= \frac{1.147}{2.94} = 0.390\ \text{m (to 3 sf).}\end{aligned}$$

Worked example 7.14

A sphere, of mass 200 g, is attached to one end of an elastic string. The other end of the string is fixed to the point O. The string has natural length 50 cm and modulus of elasticity 4.9 N. The sphere is released from rest at O and falls vertically.

(a) Calculate the maximum distance between the sphere and O.

(b) Determine the maximum speed of the sphere.

Solution

(a) Initially the sphere will only have gravitational potential energy. Let e be equal to the extension of the string and assume that at its lowest point the sphere has no gravitational potential energy. Then the gravitational potential energy lost as the sphere falls is given by

$$0.2 \times 9.8 \times (0.5 + e) = 0.98 + 1.96e$$

When the particle comes to rest it has no kinetic energy, no potential energy and so has only elastic potential energy. So at its lowest point

$$\text{EPE} = \frac{4.9e^2}{2 \times 0.5}$$
$$= 4.9e^2$$

As energy is conserved the final EPE will be equal to the initial gravitational potential energy lost, so

$$0.98 + 1.96e = 4.9e^2$$
$$0 = 4.9e^2 - 1.96e - 0.98$$

Solving this quadratic equation gives two values of e

$$e = 0.690 \text{ m or } e = -0.290 \text{ m (to 3 sf)}$$

As the second of these does not apply, because $e \geq 0$, there is a maximum extension of 0.690 m.

The maximum distance between the sphere and the point of suspension is then

$$0.690 + 0.5 = 1.190 \text{ m}$$

(b) The sphere will reach its maximum speed when its acceleration becomes zero. This will happen when the sphere reaches its equilibrium position. After this the sphere will decelerate and slow down until it comes to rest.

First find the equilibrium position. If e is the extension of the spring at the equilibrium position, then

$$0.2 \times 9.8 = \frac{4.9e}{0.5}$$
$$e = \frac{0.98}{4.9} = 0.2 \text{ m}$$

Using this value for the extension of the string and v for the speed of the sphere

$$\text{EPE} = \frac{4.9 \times 0.2^2}{2 \times 0.5}$$
$$= 0.196\text{ J}$$
$$\text{KE} = \tfrac{1}{2} \times 0.2v^2$$
$$= 0.1v^2$$
$$\text{GPE lost} = 0.2 \times 9.8 \times (0.5 + 0.2)$$
$$= 1.372\text{ J}$$

As the total energy will remain equal to the initial energy of the system

$$1.372 = 0.196 + 0.1v^2$$
$$1.176 = 0.1v^2$$

Then solving for v gives

$$v = \sqrt{11.76}$$
$$= 3.43\text{ m s}^{-1}\text{ (to 3 sf)}$$

EXERCISE 7E

1 An elastic spring has natural length 2.5 m and modulus of elasticity 100 N. Calculate the work done in extending it

(a) from 2.5 m to 2.7 m,

(b) from 2.7 m to 2.9 m.

2 In a horizontal pinball machine the spring, which has natural length 20 cm, is compressed 5 cm. If the mass of the ball is 20 g and the modulus of elasticity of the spring is 80 N. What is the speed of the ball when it leaves the spring assuming that friction can be neglected?

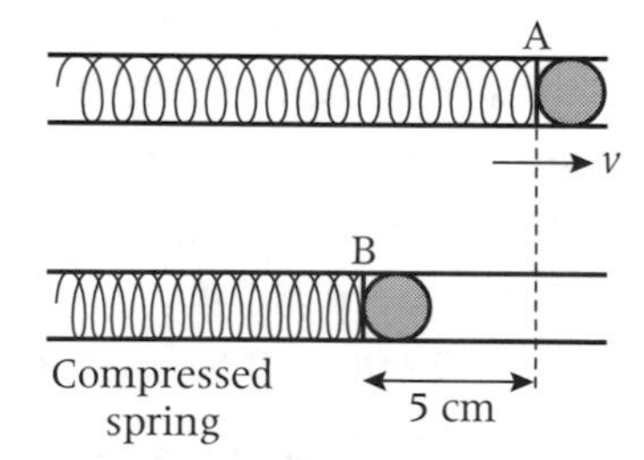

3 A 100 g mass is attached to the end B of an elastic string AB with modulus of elasticity of 3.92 N and natural length 0.25 m, the end A being fixed. The mass is pulled down from A until AB is 0.5 m and then released.

Find the velocity of the mass when the string first becomes slack and show that the mass comes to rest when it reaches A.

4 A 100 g mass is attached to the end of an elastic string which hangs vertically with the other end fixed.

The string has a modulus of elasticity of 1.96 N and natural length 0.25 m. If the mass is pulled downwards until the length of the string is 0.5 m and released, show that the mass comes to rest when the string becomes slack.

5 The mass in Question **4** is replaced by a 75 g mass. Find

(a) the velocity of the mass when the string first becomes slack,

(b) the distance below the fixed point where the mass comes to rest again.

6 The 100 g mass in Question **4** is now released from rest at the point where the string is fixed. Find the extension of the string when the mass first comes to rest. Find the speed of the mass when the string becomes slack.

7 A mass m is attached to one end B of an elastic string AB of natural length l. The end A of the string is fixed and the mass falls vertically from rest at A. In the subsequent motion, the greatest depth of the mass below A is $3l$. Calculate the modulus of elasticity of the string.

8 An energy-absorbing car bumper is modelled as a spring of natural length 20 cm and modulus of elasticity 105 kN. The 1200 kg car approaches a massive wall with a speed of 8 kph. Modelling the car as the mass spring system shown in the diagram, determine

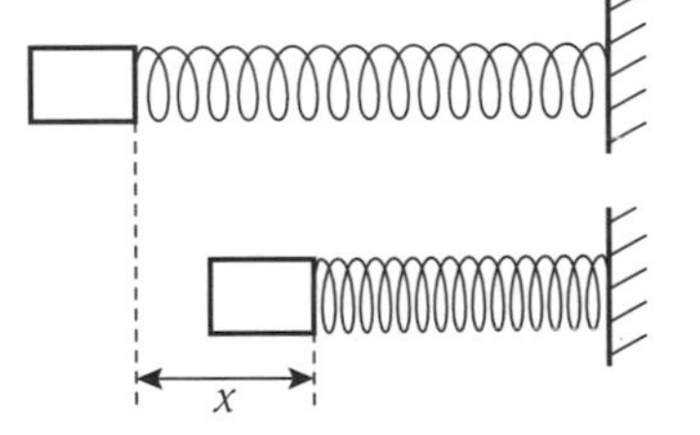

(a) the velocity of the car during contact with the wall when the spring is compressed a distance x m,

(b) the maximum compression of the spring.

9 A child's toy consists of a sphere mounted on top of a light spring, inside a smooth tube. The sphere is pushed down, so that the spring is compressed.When released, the sphere moves vertically upwards. The toy is shown in the diagram.

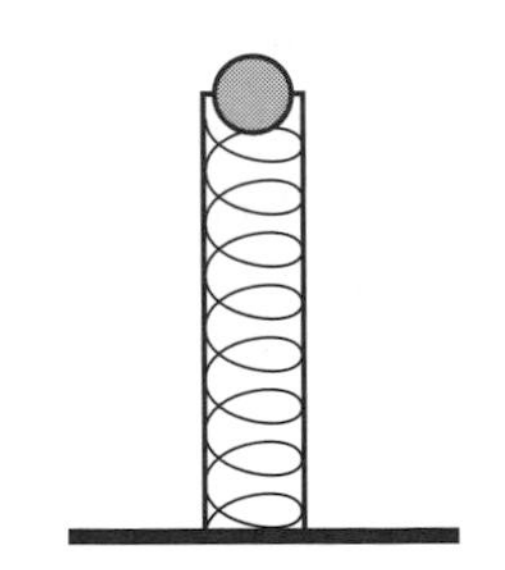

Model the sphere as a particle of mass 30 g. The spring has a natural length of 3 cm and a modulus of elasticity of 4.5 N, and rests on a horizontal surface.

(a) Find the compression of the spring when the sphere is at rest in equilibrium.

The spring is now compressed until its length is 1 cm. The system is then released.

(b) Find the maximum height of the sphere in the subsequent motion. [A]

10 A 15 tonne wagon travelling at $3.6\ \text{m s}^{-1}$ is brought to rest by a buffer (a spring) having a natural length of 1 m and modulus of elasticity of 7.5×10^5 N. Assuming that the wagon comes into contact with the buffer smoothly without rebound, calculate the compression of the buffer if there is a constant rolling friction force of 800 N.

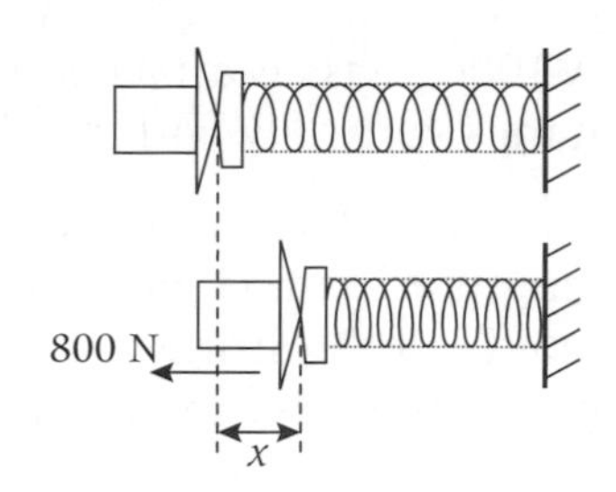

11 A 10 kg block rests on a rough horizontal table. The spring, which is not attached to the block, has a natural length of 0.8 m and modulus of elasticity 400 N. If the spring is compressed 0.2 m and then released from rest determine the velocity of the block when it has moved through 0.4 m. The coefficient of friction between the block and the table is 0.2.

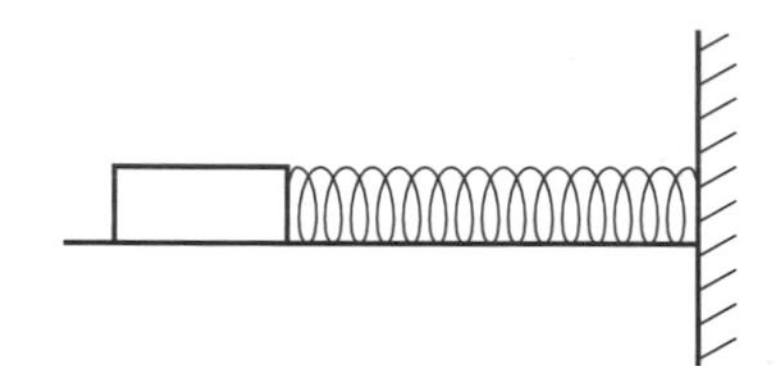

12 A platform P has negligible mass and is tied down so that the 0.4 m long cords keep the spring compressed 0.6 m when nothing is on the platform. The modulus of elasticity of the spring is 200 N. If a 2 kg block is placed on the platform and released when the platform is pushed down 0.1 m, determine the velocity with which the block leaves the platform and the maximum height it subsequently reaches.

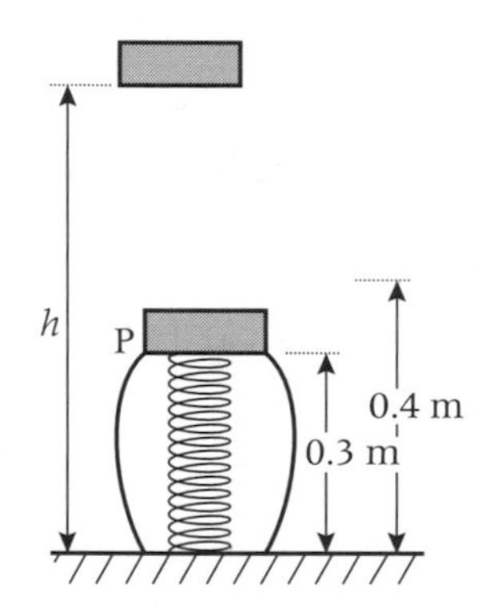

13 An archaeologist investigates the mechanics of large catapults used in sieges of castles. The diagram shows a simplified plan of such a catapult about to be fired horizontally.

The rock B of mass 20 kg is in the catapult as shown. Calculate the speed with which the rock is released at A when the elastic string returns to its natural length of 2 m if the string's modulus of elasticity is 500 N.

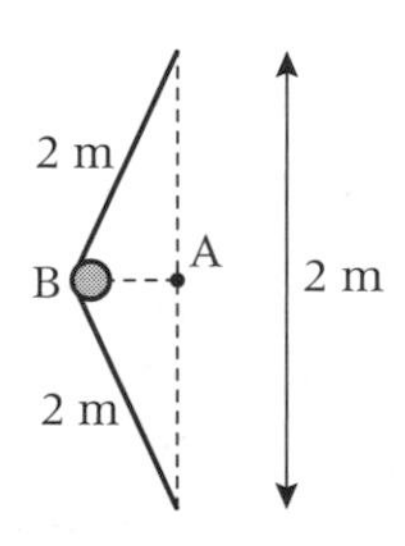

14 A ball B of mass m is attached to one end of a light elastic string of natural length a and modulus of $2mg$. The other end of the string is attached to a fixed point O. The ball is projected vertically upwards from O with speed $\sqrt{(8ga)}$; find the speed of the ball when $OB = 2a$.

Given that the string breaks when $OB = 2a$, find the speed of the ball when it returns to O. [A]

15 One end of an elastic string of modulus 20 mg and natural length a is attached to a point A on the surface of a smooth plane inclined at an angle of 30° to the horizontal. The other end is attached to a particle P of mass m. Initially P is held at rest A and then released so that it slides down a line of greatest slope of the plane. By use of conservation of energy, or otherwise, show that the speed v of P when AP $= x$, $(x > a)$, is given by

$$v^2 = \frac{g}{a}(41ax - 20a^2 - 20x^2).$$

(a) Find the maximum value of v in the subsequent motion.

(b) Find the maximum value of x in the subsequent motion. [A]

7.6 Newton's universal law of gravitation

From his study of the motion of the planets Newton formulated his law of universal gravitation. This law states that if there are two bodies of mass M and m, whose centres are a distance d apart, then each body exerts an attractive force on the other and that the magnitude, F, of this force is given by

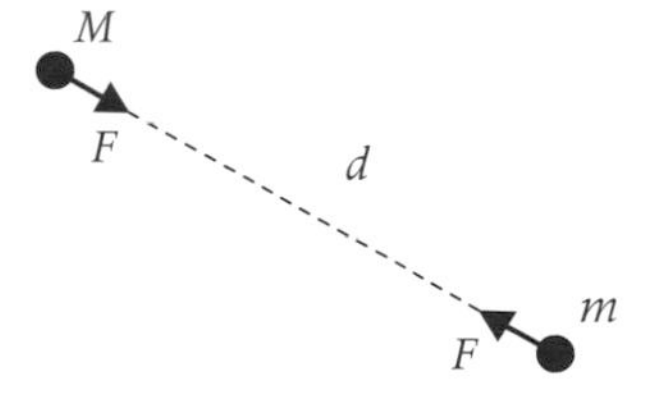

$$F = \frac{GMm}{d^2}$$

where G is a constant. The value of this constant is

$$G = 6.67 \times 10^{-11}\ \text{m}^3\ \text{kg}^{-1}\ \text{s}^{-2}.$$

Worked example 7.15

A space station, of mass 3000 kg, orbits the Earth at a height of 15 000 km above the surface of the Earth. Calculate the magnitude of the gravitational attraction on the space station.

The mass of the Earth is 5.98×10^{24} kg and the radius of the Earth is 6370 km.

Solution

The magnitude of the force is calculated using Newton's formula, with $M = 5.98 \times 10^{24}$ and $m = 3000$. The value of d is the height above the Earth plus the radius of the Earth, as it is the distance between the centre of the Earth and the space station.

$$\begin{aligned} d &= 15\,000 \times 1000 + 6370 \times 1000 \\ &= 2.137 \times 10^7 \end{aligned}$$

Substituting these values into gives

$$\begin{aligned} F &= \frac{6.67 \times 10^{-11} \times 5.98 \times 10^{24} \times 3000}{(2.137 \times 10^7)^2} \\ &= 2620\ \text{N (to 3 sf)} \end{aligned}$$

Worked example 7.16

Use Newton's universal law of gravitation to determine the value of g on the planet Mars.

The mass of Mars is 6.55×10^{23} kg and the radius of Mars is 3.36×10^6 m.

Solution

Calculate the gravitational force acting on a particle of mass m kg, taking $M = 6.55 \times 10^{23}$ and $d = 3.36 \times 10^6$.

$$F = \frac{6.67 \times 10^{-11} \times 6.55 \times 10^{23} \times m}{(3.36 \times 10^6)^2}$$

$$= 3.87m \text{ N}$$

As this must be equal to mg on Mars, we have $g = 3.87 \text{ m s}^{-2}$ on the planet Mars.

EXERCISE 7F

1 A satellite, of mass 300 kg, orbits the Earth at a height of 7000 km above the surface of the Earth. Calculate the magnitude of the gravitational attraction that acts on the satellite.

2 The mass of the Moon is 7.38×10^{22} kg and the radius of the Moon is 1.73×10^6 m. Determine the acceleration due to gravity on the Moon.

3 A planet has a mass of 5×10^{20} kg and on this planet the acceleration due to gravity is 3.2 m s^{-1}. Determine the radius of the planet.

4 A man, of mass 80 kg, climbs 5000 m to the top of a mountain.

(a) Use the universal law of gravitation to calculate the gravitational attraction on the man.

(b) Compare your answer to **(a)** with the result given by simply using mg.

5 The mass of the Sun is 1.99×10^{30} kg and the average distance between the centres of the Earth and the Sun is 1.47×10^7 km.

(a) Calculate the attractive force that the Sun exerts on the Earth.

(b) What force does the Earth exert on the Sun?

7.7 Work and the law of universal gravitation

You have used the result that the gravitational potential energy is given by mgh. This is satisfactory, but for large values of h you should treat gravity as a variable force and calculate the gravitational potential energy by integrating the expression for the gravitational attraction given in the law of universal gravitation.

For example to calculate the gravitational potential energy at a height h above the surface of the Earth you need to evaluate the integral

$$\int_r^{r+h} \frac{GMm}{x^2}\,dx$$

where r is the radius of the Earth.

Worked example 7.17

A climber, of mass 75 kg, climbs to the top of a mountain of height 2000 m above the surface of the Earth. Calculate the gravitational potential energy of the climber relative to the surface of the Earth.

Note that the radius of the Earth is 6.37×10^6 m and that its mass is 5.98×10^{24} kg. Also G is $6.67 \times 10^{-11}\,\text{m}^3\,\text{kg}^{-1}\,\text{s}^{-2}$.

Solution

The gravitational potential energy is given by

$$\int_r^{r+h} \frac{GMm}{x^2}\,dx = \left[-\frac{GMm}{x}\right]_r^{r+h}$$

$$= GMm\left(\frac{-1}{r+h} + \frac{1}{r}\right)$$

$$= \frac{GMmh}{r(r+h)}$$

Substituting the values of G, M, m, r and h gives

$$\frac{6.67 \times 10^{-11} \times 5.98 \times 10^{24} \times 75 \times 2000}{6.37 \times 10^6(6.37 \times 10^6 + 2000)} = 1\,470\,000\text{ J (to 3 sf)}$$

Note that $75 \times 9.8 \times 2000 = 1\,470\,000$ J (to 3 sf), so at this height there is no real need to consider gravity as a variable force.

Worked example 7.18

A satellite, of mass 400 kg, is to orbit the Earth at a height of 3.6×10^7 m above the surface, where it is to travel at $230\,\text{m}\,\text{s}^{-1}$. Find the work that must be done to position the satellite in this way.

7

Solution

First calculate the kinetic energy of the satellite.

$$\tfrac{1}{2} \times 400 \times 230^2 = 1.06 \times 10^7\text{ J (to 3 sf)}$$

Then consider the work done against gravity to raise the satellite to this height. We have the result below from the previous worked example.

$$\int_r^{r+h} \frac{GMm}{x^2}\,dx = \frac{GMmh}{r(r+h)}$$

Substituting $G = 6.67 \times 10^{-11}$, $M = 5.98 \times 10^{24}$, $m = 400$, $r = 6.37 \times 10^6$ and $h = 3.6 \times 10^7$, gives

$$\frac{6.67 \times 10^{-11} \times 5.98 \times 10^{24} \times 400 \times 3.6 \times 10^7}{6.37 \times 10^6(6.37 \times 10^6 + 3.6 \times 10^7)} = 2.13 \times 10^{10}\text{ J (to 3 sf)}$$

Total work done $= 1.06 \times 10^7 + 6.38 \times 10^{10} = 2.13 \times 10^{10}$ J (to 3 sf).

EXERCISE 7G

1 A rocket rises to a height of 4×10^8 m above the surface of the Earth. Assume that the mass of the rocket remains constant at 500 kg. Calculate the gravitational potential energy of the rocket at this height relative to the surface of the Earth.

2 A climber, of mass 85 kg, reaches the top of Mount Everest at a height of 8850 m above sea level.

(a) Calculate the gravitational potential energy of the climber relative to sea level.

(b) Would it be reasonable to calculate the gravitational potential energy using mgh at this altitude?

3 A satellite, of mass 600 kg, is to be put into orbit at a height of 4.5×10^5 m above the surface of the Earth, where it should travel at 300 m s^{-1}. Calculate the work that must be done to put the satellite into orbit in this way.

4 A mountain range on the planet Mars has a peak at a height of 5000 m above the surface of the planet. Find the gravitational potential energy, relative to the surface of the planet, of an alien who is on this peak. The mass of the alien is 120 kg.

Note: the mass of Mars is 6.55×10^{23} kg and the radius of Mars is 3.36×10^6 m.

5 The work done for an object to escape from the Earth's gravitational influence can be calculated by assuming that the object reaches an infinite height. Calculate the work done in this way for a rocket of mass 2000 kg.

7.8 Power

Power is defined as the rate of doing work. For example we may talk about a more powerful car or motorbike. This is a way of describing how quickly they gain kinetic energy. The more quickly kinetic energy is gained, the shorter the time that the work is done in and so the more powerful the vehicle.

A simple definition of power is:

$$\text{Power} = \frac{\text{work done}}{\text{time taken}}.$$

Worked example 7.19

Hannah, who has mass 50 kg, climbs a flight of stairs in 20 s. As she climbs the stairs she rises a total of 4 m.

(a) Calculate the work done as she climbs the stairs.

(b) Calculate her power output as she climbs the stairs.

Solution

(a) Work done $= 50 \times 9.8 \times 4$

$= 1960\text{ J}$

(b) Power $= \dfrac{\text{work done}}{\text{time taken}}$

$= \dfrac{1960}{20}$

$= 98\text{ W}$

Note that the SI unit for power is the watt (W). An alternative unit would be J s^{-1}. Another, more traditional, unit that is sometimes used for power is the horsepower (hp), and is such that 1 hp is approximately 740 watts.

Work done by force

An alternative approach to power is to derive a formula based on the definition that power is the rate of doing work.

The work done by a force is Fs, where F is the magnitude of the force and s is the displacement. As rates can found by differentiating with respect to t, we have

$$\text{Power} = \frac{d}{dt}(Fs).$$

Considering the case of a constant force leads to

$$\begin{aligned}\text{Power} &= \frac{d}{dt}(Fs)\\ &= F\frac{ds}{dt}\\ &= Fv\end{aligned}$$

This result is very useful and can be applied in many examples.

Worked example 7.20

A car experiences a resistance force of magnitude 1200 N, when travelling at a constant speed of 25 m s^{-1}. Calculate the power output of the car.

Solution

As the car is travelling at a constant speed, there must be a forward force on the car equal in magnitude to the resistance force. So in this case the force F exerted by the car has magnitude 1200 N and $v = 25$ m s^{-1}. Using $P = Fv$, gives

$$\begin{aligned} P &= 1200 \times 25 \\ &= 30\,000 \text{ W} \end{aligned}$$

Worked example 7.21

A car, of mass 1200 kg, experiences a resistance force that is proportional to its speed. The car has a maximum power output of 36 000 W and a maximum speed of 40 m s^{-1}.

(a) Determine an expression for the magnitude of the resistance force, when the speed of the car is v m s^{-1}.

(b) Find the power output if the car is accelerating at 2 m s^{-2} and is travelling at 10 m s^{-1}.

(c) Calculate the maximum acceleration of the car when it is travelling at 20 m s^{-1}.

Solution

(a) As the resistance is proportional to the speed we have

$$R = kv.$$

At its top speed the resistance force will be equal in magnitude to the forward force exerted by the car. Using $P = Fv$ gives

$$\begin{aligned} 36\,000 &= 40k \times 40 \\ k &= \frac{36\,000}{40^2} \\ &= 22.5 \end{aligned}$$

(b) The car exerts a forward force, of magnitude F N, and experiences a resistive force of magnitude 22.5×10 N. So the resultant force on the car is $F - 225$. As the car is accelerating at 2 m s^{-2}, you can apply Newton's second law to give

$$\begin{aligned} F - 225 &= 1200 \times 2 \\ F &= 2625 \text{ N} \end{aligned}$$

Now the power output can be found using $P = Fv$ as

$$\begin{aligned} P &= 2625 \times 10 \\ &= 26\,250 \text{ W} \end{aligned}$$

(c) At a speed of $20\ \text{m s}^{-1}$, the car will experience a resistance force of 450 N. If it exerts a forward force of magnitude F N and accelerates at $a\ \text{m s}^{-2}$, then Newton's second law can be applied to give

$$F - 450 = 1200a$$
$$F = 1200a + 450$$

Now using the relationship $P = Fv$, you can form and solve an expression for a.

$$36\,000 = (1200a + 450) \times 20$$
$$a = \frac{1800 - 450}{1200}$$
$$= 1.125\ \text{m s}^{-2}$$

Worked example 7.22

A car has a maximum power of 32 000 W and a mass of 1200 kg. The resistance forces acting on the car are modelled as being proportional to the speed of the car. The car has a top speed of $40\ \text{m s}^{-1}$.

(a) Find the constant of proportionality in the model for the resistance forces.

(b) Find the maximum speed of the car up a slope that is inclined at an angle α to the horizontal, where $\sin \alpha = \frac{1}{14}$.

(c) Find the power output of the car if it is accelerating at $2\ \text{m s}^{-2}$, while travelling at $5\ \text{m s}^{-1}$, on a horizontal road.

Solution

(a) At top speed the resistance forces R are equal to the forward forces F. Using $P = Fv$ and assuming that $R = kv$ gives

$$32\,000 = k \times 40^2$$
$$k = \frac{32\,000}{40^2}$$
$$= 20$$

(b) When travelling up the hill at a constant speed, the resultant force must balance both the component of gravity down the slope and the resistance force. So the force, F, exerted by the car is given by

$$F = mg \sin \alpha + kv$$
$$= 1200 \times 9.8 \times \tfrac{1}{14} + 20v$$
$$= 840 + 20v$$

Using $P = Fv$ gives

$$32\,000 = (840 + 20v)v$$
$$20v^2 + 840v - 32\,000 = 0$$
$$v^2 + 42v - 1600 = 0$$

Solving this quadratic gives:

$$v = \frac{-42 \pm \sqrt{42^2 - 4 \times 1 \times (-1600)}}{2 \times 1}$$
$$= 24.2 \text{ or } -66.2 \text{ m s}^{-1}$$

So the maximum speed up the slope is 24.2 m s^{-1}.

(c) If the car is travelling at 5 m s^{-1}, then there will be a resistance force of 100 N:

$$F - 100 = 1200 \times 2$$
$$= 2500 \text{ N}$$

The power can then be found using $P = Fv$:

$$P = 2500 \times 5$$
$$= 12\,500 \text{ W}$$

EXERCISE 7H

1 A crane lifts a load, of mass 800 kg, through a height of 12 m in 2 minutes.

(a) Calculate the work done by the crane.

(b) Find the power of the crane.

2 A child, of mass 56 kg, climbs up a flight of stairs in 49 s. There are 50 steps, each of height 18 cm. Calculate the rate at which the child was working as she climbed the stairs.

3 A train travels at a constant speed of 30 m s^{-1} and experiences a resistance force of magnitude 30 000 N at this speed. Calculate the power output of the train.

4 A pump is used to raise water from ground level into a tank at a height of 5m. The pump is able to pump 5000 litres per hour. Find the power of the pump. (The mass of 1 litre of water is 1 kg.)

5 A car, of mass 1000 kg, has a maximum power output of 36 000 W and a maximum speed of 40 m s^{-1}. The resistance force on the car is proportional to its speed.

(a) Find the magnitude of the resistance force when it is travelling at its maximum speed.

(b) Find the magnitude of the resistance force when the car is travelling at 30 m s^{-1}.

(c) Find the maximum possible acceleration of the car when it is travelling at 30 m s^{-1}.

6 A motorcycle, of mass 300 kg, has a maximum power output of 30 kW and a top speed of 40 m s^{-1} on the horizontal. Assume that the resistance forces that act on the motorcycle are proportional to its speed.

(a) Find an expression for the magnitude of the resistance force acting on the motorcycle when it is travelling at v m s^{-1}.

(b) Calculate the power output of the motorcycle if it travels at a constant 25 m s^{-1}.

(c) Find the maximum possible speed of the motorcycle, while it is accelerating at 2 m s^{-2}.

7 A car, of mass 1000 kg, is assumed to experience a resistance force that is proportional to its speed. The car has a maximum power output of 50 000 W and a top speed of 50 m s^{-1} on the horizontal. A slope is inclined at an angle α to the horizontal, where $\sin \alpha = \frac{1}{10}$. Find the maximum speed of the car when travelling up or down the slope.

8 A motorcycle, of mass 300 kg, has a maximum power output of 30 kW and a top speed of 45 m s^{-1} on the horizontal. Assume that the resistance forces that act on the motorcycle are proportional to its speed.

(a) Find an expression for the magnitude of the resistance force that acts on the motorcycle at a speed of v m s^{-1}.

(b) Find the maximum possible speed of the motorcycle as it travels up a slope inclined at 5° to the horizontal.

(c) What would be the maximum speed of the motorcycle down the same slope ?

9 A car, of mass 1000 kg, has a power output of 24 000 W. It experiences a resistance force of magnitude $20v$ when travelling at a speed v.

(a) Find the maximum speed of the car on a horizontal road.

(b) Find the maximum speed of the car when travelling up a slope inclined at an angle α to the horizontal, where $\sin \alpha = \frac{1}{20}$.

10 A car, of mass 1000 kg, is assumed to experience a resistance force that is proportional to its speed squared. The car has a maximum power output of 32 000 W and a top speed of 40 m s^{-1}, on the horizontal.

(a) Find the resistance force acting when the car travels at a speed of 20 m s^{-1}.

(b) The car travels 500 m up a slope inclined at an angle α to the horizontal, where $\sin \alpha = \frac{1}{15}$. The car travels at a constant speed of 20 m s^{-1}. Find the work done by the car as it travels up the slope.

11 A cyclist can pedal up a slope, inclined at 4° to the horizontal, at a maximum speed of 2 m s^{-1}. Model the cycle as a particle of mass 70 kg. Assume that there are no resistance forces acting on the cyclist.

(a) When he is pedalling up the slope at his maximum speed, show that the power output of the cyclist is approximately 96 W.

(b) If the power output remains the same, find the maximum speed of the cyclist when travelling up a slope inclined at 6° to the horizontal.

(c) **(i)** When modelling the motion uphill, explain why it is reasonable to assume that there is no resistance.

(ii) When trying to find the maximum speed of the cyclist down the slope, explain why it is **not** reasonable to assume that there is no resistance. [A]

12 A car, of mass 1200 kg, has a maximum power output of 48 000 W. On a horizontal road the car has a maximum speed of 40 m s^{-1}. Assume that the resistance forces acting on the car are proportional to its speed.

(a) Find the resistance force acting on the car when it travels at v m s^{-1}.

(b) Find the percentage reduction in the power output of the car if its speed is reduced by 10%.

(c) Use your answer to **(b)** to describe one advantage of reducing the speed at which the car is driven.

(d) Find the maximum speed of the car, when being driven up a slope at 4° to the horizontal. [A]

13 The maximum power output of a car is 50 000 W and its top speed, on a horizontal road, is 40 m s^{-1}. In order to model the motion of the car, assume that it experiences a resistance force proportional to its speed.

(a) Find the resistance force when the car is travelling at 20 m s^{-1}.

(b) When the car tows a caravan the resistance force is increased by 50%. Find the maximum speed of the car when it tows the caravan on a horizontal road. [A]

14 A car and its driver, of a total mass 500 kg, are ascending a hill of inclination $\sin^{-1}\left(\frac{1}{7}\right)$ to the horizontal with a constant speed of 5 m s^{-1}. Given that the motion is opposed by a frictional force of magnitude 800 N, find the power generated by the engine of the car.

The driver presses the accelerator, which has the effect of suddenly increasing the power to 20 kW. Calculate the resulting acceleration of the car. [A]

15 A car, of mass 1000 kg, travels up a hill inclined at 4° to the horizontal. Assume that the car experiences a constant resistance force, and is moving at constant or increasing speed.

(a) Draw and label a diagram to show the forces acting on the car, if it is modelled as a particle. Describe one weakness of modelling the car as a particle.

(b) If the car exerts a forward force of 1000 N when travelling at a constant speed of 20 m s^{-1} up the hill, show that the magnitude of the resistance force is 316 N to the nearest newton.

(c) Use your answer to **(b)** to find the power output of the car if it is travelling at 20 m s^{-1} and accelerating at 1.5 m s^{-2}. [A]

Key point summary

1 The kinetic energy of a body is defined as $\frac{1}{2}mv^2$. p109

2 The work done by the force is the quantity Fs. p111

3 Gravitational potential energy $= mgh$. p117

4 Hooke's law states that the tension T, in a spring, is given by $\frac{\lambda x}{l}$. p123

5 The elastic potential energy (EPE) of a stretched or compressed spring $= \frac{\lambda e^2}{2l}$. p126

6 Power $= \frac{\text{work done}}{\text{time taken}} =$ the rate of doing work. p135

Formulae to learn

Kinetic energy $\frac{1}{2}mv^2$

Work done Fs or $Fs \cos \theta$

Work done = change in kinetic energy

Gravitational potential energy mgh

Hooke's law $T = \frac{\lambda x}{l}$

Work done by a variable force $\int f(x)dx$

Elastic potential energy $\frac{\lambda e^2}{2l}$

Power $P = Fv$

Test yourself	What to review
1 Calculate the gain in kinetic energy as a car, of mass 1200 kg, increases in speed from 10 to 25 m s^{-1}.	*Section 7.1*
2 A force of 500 N acts on a car, of mass 1250 kg, in the direction of motion. The car has an initial speed of 8 m s^{-1}. The force acts as the car travels 25 m on a horizontal surface. **(a)** Calculate the work done by the force as the car travels this distance. **(b)** Find the final speed of the car.	*Section 7.2*
3 A go-kart, of mass 50 kg, is at rest at the top of a slope inclined at 8° to the horizontal. Find the speed of the go-kart when it has travelled 70 m down the slope **(a)** if there is no resistance to the motion, **(b)** there is a constant 20 N resistance force.	*Section 7.3*

4 A 'dropslide' at a leisure park consists of a curved section AB and a horizontal section BC shown below. Children start at rest at the point A and slide down to B and on towards C. The points, A, B and C all lie in the same vertical plane and the motion of the child is in this plane. *Section 7.3*

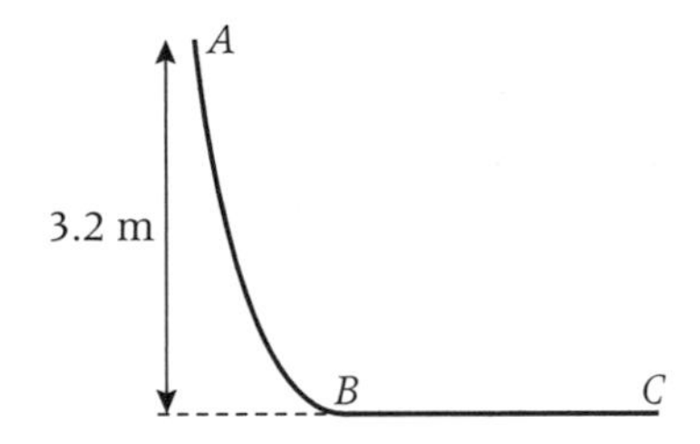

A child of mass 30 kg uses the slide.

(a) Assuming that there is no friction or air resistance acting on the child on the section AB, find the speed of the child at B. How would this speed compare with that of a heavier child?

(b) If the child travels 5 m along BC before stopping, find the magnitude of the friction force between the child and the surface. State two factors that would influence the magnitude of this force for different children. [A]

Test yourself (continued)	What to review
5 A sphere, of mass 500 g, is attached to an elastic string of natural length 70 cm and modulus of elasticity 70 N. One end of the string is fixed to a point O. **(a)** Find the length of the elastic when the sphere hangs in equilibrium. The sphere is released from the point O. **(b)** Find the maximum distance between the sphere and O. **(c)** Find the speed of the sphere when the length of the elastic string is 80 cm. **(d)** Find the maximum speed of the sphere.	*Section 7.5*
6 Calculate the work done against gravity in raising a rocket, of mass 400 kg, from ground level to a height of 50 000 m.	*Section 7.7*
7 A car of maximum power output 32 kW and mass 800 kg, is travelling up a slope at an angle θ to the horizontal. A simple model for the motion of the car assumes that there are no resistance forces acting on the car. **(a)** Find the angle θ, to the nearest degree, for the steepest slope that the car can ascend at a speed of 10 m s^{-1}. A refined model for the motion of the car would take account of the resistance forces on the car. **(b)** The slope is in fact at 5° to the horizontal, and the car is still travelling at its maximum speed of 10 m s^{-1}. Find the magnitude of the resistance forces on the car. **(c)** The resistance forces on the car are assumed to be proportional to the speed. Use your result to **(b)** to find a simple model for the resistance forces.	*Section 7.8* [A]

Test yourself ANSWERS

1 315 000 J.

2 **(a)** 12 500 J, **(b)** 9.17 m s^{-1}.

3 **(a)** 13.8 m s^{-1}, **(b)** 11.6 m s^{-1}.

4 **(a)** 7.92 m s^{-1}, same, **(b)** 188 N, mass, clothes.

5 **(a)** 74.9 cm, **(b)** 1.015 m, **(c)** 3.70 m s^{-1}, **(d)** 3.77 m s^{-1}.

6 1.95×10^8 J.

7 **(a)** 24°, **(b)** 2517 N, **(c)** $252v$ N.

CHAPTER 8

Circular motion

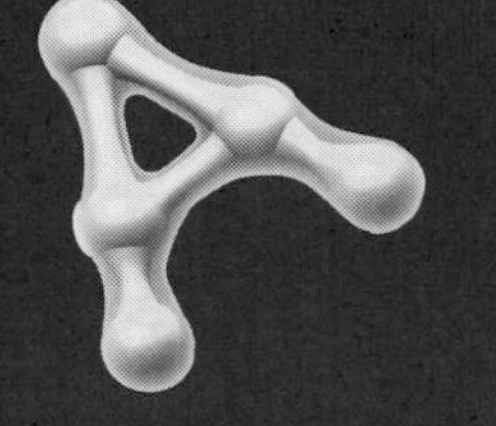

Learning objectives

After studying this chapter you should be able to:

- convert from rpm to rad s^{-1}
- know that the velocity is directed along the tangent of the circle
- know that the acceleration is directed towards the centre of the circle
- know that $v = r\omega$
- know that $a = r\omega^2 = \frac{v^2}{r}$
- solve problems for motion in horizontal or vertical circles at constant speeds.

8.1 Introduction

This chapter will focus on the motion of objects that travel in a circle with constant speed. There are many everyday situations which can be modelled in this way, for example, fairground rides, satellites orbiting the Earth and clothes in a spin-dryer. Also cars travelling around corners or on roundabouts travel round part of a circle. To analyse such situations we need to apply Newton's second law after finding the resultant force along with the acceleration of the object travelling round the circle. Before doing this you must be familiar with the concept of angular speed.

8.2 Angular speed

Imagine a line drawn from the centre of the circle to the object that is travelling round the circle. The rate at which this line rotates about the centre is called the angular speed of the object. This could be given in terms of rpm (revolutions per minute), as with car engines, but it is most often given in radians per second (rad s^{-1}).

We will now consider an example that makes use of angular speed. To do this you will need to recall that one complete revolution is equivalent to turning through 2π radians.

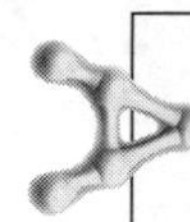

$$1 \text{ rpm} = \frac{2\pi}{60} \text{ rad s}^{-1}$$

Worked example 8.1

The angular speed of a record is given as $33\frac{1}{3}$ rpm. Find the angular speed of the record in radians per second.

Solution

In one revolution there are 2π radians.

$$\begin{aligned}\text{So } 33\tfrac{1}{3} \text{ rpm} &= 33\tfrac{1}{3} \times 2\pi \text{ radians per minute}\\ &= \frac{33\frac{1}{3} \times 2\pi}{60} \text{ radians per second}\\ &= \frac{10\pi}{9}\\ &= 3.49 \text{ rad s}^{-1}\end{aligned}$$

Velocity and circular motion

Consider now a particle, which moves with angular speed ω rad s^{-1}. The angle turned through in 1 s will be ω radians, so the angle turned through in t s will be $\theta = \omega t$ radians. If the particle moves in a circle with radius r m, the distance moved by the particle will be the length of the arc AB.

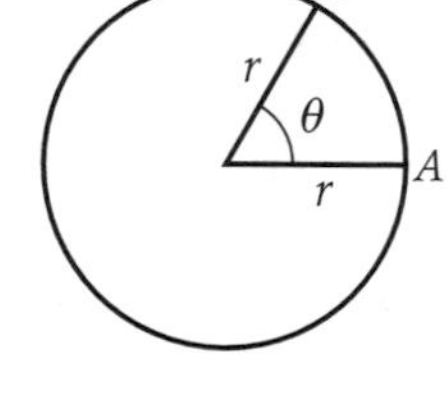

$$AB = r\theta$$

Each second the particle moves a distance $r\omega$, so the speed, v, of the particle is given by

$$v = r\omega$$

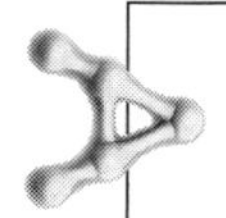

The velocity has magnitude $r\omega$ and is directed along a tangent to the circle.

Acceleration and circular motion

Newton's first law of motion states that a particle will move in a straight line unless acted upon by a force. Therefore, if a particle moves in a circle, there must be a force acting upon it. If the speed is constant then the resultant force will have no component in the direction of motion. Hence the resultant force on the particle will be perpendicular to the velocity and must always act towards the centre. Consequently, if the resultant force is towards the centre, then so is the acceleration of the particle. You will now find the magnitude of this acceleration.

Suppose a particle moves in a circle with centre O, radius r, and with constant angular velocity ω.
Suppose further that the particle starts from A($t = 0$) so $\theta = \omega t$.
The position vector of the particle **OP**, in terms of the unit vectors **i** and **j** will be given by

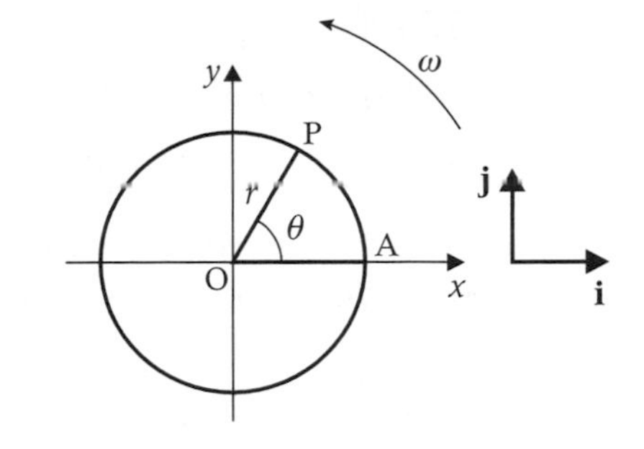

$$\begin{aligned}\mathbf{r} &= x\mathbf{i} + y\mathbf{j}\\ &= r\cos\theta\,\mathbf{i} + r\sin\theta\,\mathbf{j}\\ &= r\cos\omega t\,\mathbf{i} + r\sin\omega t\,\mathbf{j}\end{aligned}$$

If you differentiate once you get the velocity vector and twice you get the acceleration vector.

$$\mathbf{v} = -\omega r\sin\omega t\,\mathbf{i} + \omega r\cos\omega t\,\mathbf{j}$$

and

$$\mathbf{a} = -\omega^2 r\cos\omega t\,\mathbf{i} - \omega^2 r\sin\omega t\,\mathbf{j}$$

The magnitude of the acceleration of the particle is:

$$a = \omega^2 r\sqrt{\cos^2\omega t + \sin^2\omega t}$$

But $\cos^2\theta + \sin^2\theta \equiv 1$, so

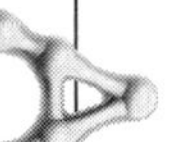

$$a = r\omega^2$$

Sometimes it is useful to express the magnitude of the acceleration in terms of the speed, v, rather than ω. From $v = r\omega$, note that $\omega = \dfrac{v}{r}$ and substitute this so that the expression for the magnitude of the acceleration becomes

$$a = r\left(\frac{v}{r}\right)^2 = \frac{v^2}{r}$$

To confirm that the direction of the particle's acceleration is towards the centre compare the acceleration and position vectors.

$$\begin{aligned}\mathbf{a} &= -\omega^2(r\cos\omega t\,\mathbf{i} + r\sin\omega t\,\mathbf{j})\\ \mathbf{a} &= -\omega^2\mathbf{r}\end{aligned}$$

The position vector is directed outwards from the centre of the circle. The acceleration is in the opposite direction and so is directed **towards** the centre.

Worked example 8.2

If the radius of the Earth is taken as 6370 km find the speed and magnitude of the acceleration of a man standing on the Equator.

$$\begin{aligned}\omega &= 1 \text{ rev per day}\\ &= \frac{2\pi}{24\times 60\times 60} = 7.27\times 10^{-5}\text{ rad s}^{-1}\end{aligned}$$

$$v = 6\,370\,000 \times 7.27\times 10^{-5}\text{ m s}^{-1} = 463\text{ m s}^{-1}$$

$$a = 637\,000 \times (7.27\times 10^{-5})^2 = 3.37\times 10^{-2}\text{ m s}^{-2}$$

EXERCISE 8A

1 What is the angular speed of the minute hand of a clock in

(a) revolutions per minute, (b) radians per second.

2 A wheel makes 100 revolutions in 10 minutes. Find its angular speed in radians per second.

3 The distance of the Moon from the Earth is approximately 355 000 km. Estimate the speed of the Moon relative to the Earth in $m\,s^{-1}$. (Assume that the Moon rotates round the Earth twice a day in a circular orbit.)

4 The distance from the Sun to the Earth is approximately 150×10^5 km. Estimate the speed of the Earth relative to the Sun. (Assume that the orbit of the Earth is a circle and that the Earth orbits the Sun once every year.)

5 Find the speed and the magnitude of the acceleration of a particle, which moves in a circle with radius 20 cm, and angular speed of 2500 rpm.

6 Find the magnitude of the resultant force on a particle of mass 250 g, which moves in a circle of radius 10 m and angular speed of 36 rpm.

7 A particle moves in a circle with speed $5\,m\,s^{-1}$. Given that the acceleration has magnitude $10\,m\,s^{-2}$, find the radius of the circle.

8 What is the magnitude of the acceleration of a truck, which goes round a bend of radius 20 m at a speed of $20\,km\,h^{-1}$.

9 A washing machine spins at 1000 rpm. The drum has diameter 40 cm. What are the speed and magnitude of the acceleration of a sock, which is stuck to the edge of the drum during the spinning cycle?

10 Joe and Tom ride on a fairground roundabout. Joe is 2 m and Tom is 1.5 m from the centre of rotation and the roundabout is rotating at 10 rpm. Find:

(a) the angular speed of the roundabout in $rad\,s^{-1}$,

(b) the speeds of Joe and Tom in $m\,s^{-1}$.

8.3 Forces involved in horizontal circular motion

In the previous section you saw that when a particle, of mass m, moves in a circle with constant speed, its acceleration is of magnitude $r\omega^2$ (or v^2/r) and acts towards the centre of the circle (where r is the radius of the circle, v is the speed of the particle and ω is its angular speed). The resultant force, therefore, must always act towards the centre. Newton's second law implies that the magnitude of the resultant force will be given by

$$F = mr\omega^2 \qquad (\text{or } mv^2/r)$$

If the circle of motion is horizontal then any vertical components of force must cancel.

When a particle moves in a horizontal circle with constant speed there are two principles that can always be applied:

- resultant of vertical components of forces must be zero
- $F = ma$ can be applied radially.

Worked example 8.3

A particle P, of mass 100 g, is attached to one end of a light inextensible string of length 50 cm. The other end of the string is fixed at O, on a smooth horizontal surface. The particle moves in a circle, with centre O, at 180 rpm. Find the tension in the string.

Solution

The diagram shows the forces acting on the particle, its weight, the normal reaction from the surface and the tension in the string.

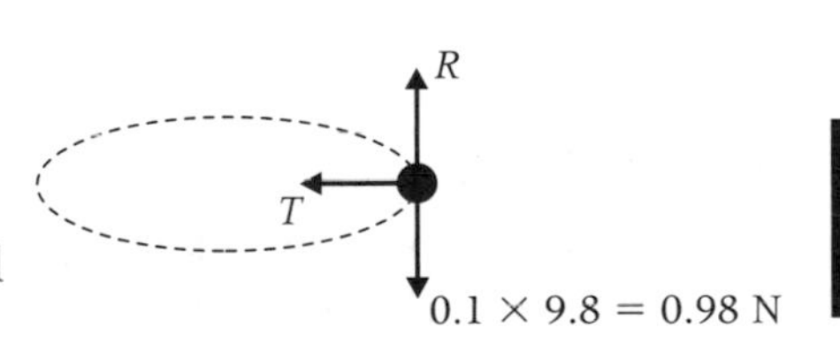

The acceleration of P is horizontal, so the resultant of the vertical forces must be zero.

$$R = 0.98 \text{ N}$$

The angular speed must be converted from rpm to rad s^{-1}.

$$\begin{aligned}\omega &= 180 \text{ rpm}\\ &= 180 \times \frac{2\pi}{60} \text{ rad s}^{-1}\\ &= 6\pi \text{ rad s}^{-1}\end{aligned}$$

Using $F = ma$ in the radial direction gives

$$\begin{aligned}T &= 0.1 \times 0.5 \times (6\pi)^2\\ &= 17.8 \text{ N}\end{aligned}$$

Worked example 8.4

A turntable rotates at $33\frac{1}{3}$ rpm. A counter, of mass 10 g, is placed 10 cm from the centre of the circle and does not slip. Find the minimum possible value of the coefficient of friction.

Solution

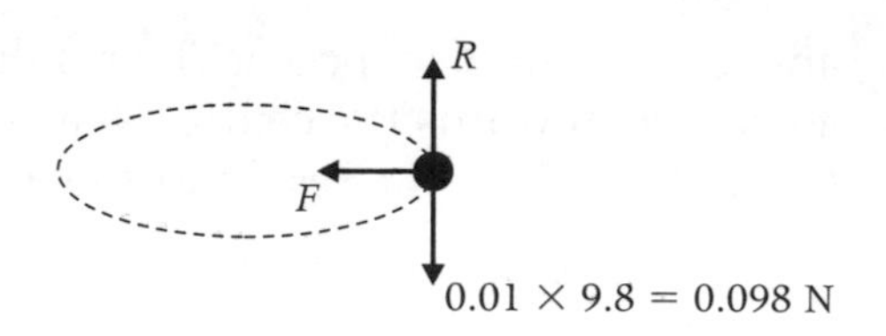

The diagram shows the forces acting on the counter.

First express the angular speed in rad s^{-1}.

$$\omega = \frac{33\frac{1}{3} \times 2\pi}{60}$$

$$= \frac{10\pi}{9} \text{ rad s}^{-1}$$

Using $F = ma$ radially with $a = r\omega^2$

$$F = 0.01 \times 0.1 \left(\frac{10\pi}{9}\right)^2$$

$$= 0.01218 \text{ N}$$

Resolving vertically

$$R = 0.01 \times 9.8 = 0.098 \text{ N}$$

Then we can use the friction inequality $F \leqslant \mu R$, to give

$$0.01218 \leqslant \mu \times 0.098$$

$$\mu \geqslant 0.124 \text{ (to 3 sf)}$$

So the least value of μ is 0.124.

EXERCISE 8B

1 A particle, of mass 2 kg, is attached to a fixed point on a smooth horizontal table by a light inextensible string of length 50 cm. The particle travels in a circle on the table at 400 rpm. Find the tension in the string.

2 A particle, of mass 2 kg, is moving in a circle of radius 5 m with a constant speed of 3 m s^{-1}.What is the magnitude and direction of the resultant force acting on the particle?

3 A particle, of mass 3 kg, moves in a circle on a smooth horizontal plane. The particle is attached to a fixed point, O, on the plane, by a light inextensible string of length 1.5 m. If the velocity of the particle is 6 m s^{-1}, find the tension in the string.

4 An inextensible string has length 3 m. It is fixed at one end to a point O on a smooth horizontal table. A particle of mass 2 kg is attached to the other end and describes circles on the table with O as centre and the string taut. If the string breaks when the tension is 90 N, what is the maximum safe speed of the particle?

5 A marble is made to rotate against the outside rim of a circular tray of radius 0.2 m. The mass of the marble is 100 g and it moves at 2 m s^{-1}. Calculate the horizontal force that the tray exerts on the marble.

6 An athlete throwing the hammer swings it in preparation for his throw. Assume that the hammer travels in a horizontal circle of radius 2.0 m and is rotating at 1 revolution per second. The mass of the wire is negligible and the mass of the hammer is 7.3 kg. What is the tension in the wire attached to the hammer? (Ignore the weight of the hammer.)

7 A particle moves with constant angular speed around a circle of radius a and centre O. The only force acting on the particle is directed towards O and is of magnitude $\frac{k}{a^2}$ per unit mass, where k is a constant. Find, in terms of k and a, the time taken for the particle to make one complete revolution.

8 A satellite, of mass 1 tonne, orbits a planet at 1 revolution per day. The satellite is at a height of 700 km above the surface of the planet. The radius of the planet is 6500 km. Find the force of gravity on the satellite.

9 A car, of mass 1 tonne, takes a bend, of radius 150 m, on a level road, at 80 km h^{-1}, without sliding. Find the frictional force between the tyres and the ground. What is the least value of μ?

10 Two particles, P and Q, are connected by a light inextensible string, that passes through a hole in a smooth horizontal table. Particle P has mass m and travels in a circle, whereas Q hangs in equilibrium under the table and has mass $2m$. If the radius of the circle in which P moves is 50 cm, find the angular velocity of P.

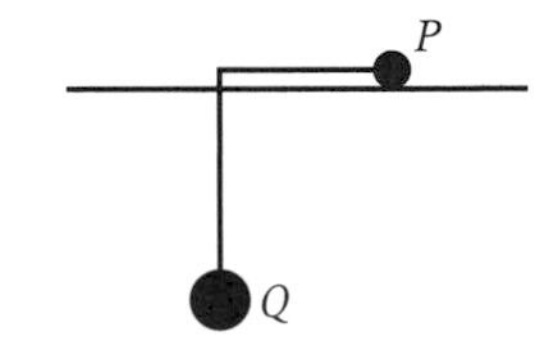

11 A horizontal turntable rotates at 45 rpm. A coin, of mass 50 g, is placed on the turntable, at a distance 20 cm from the centre of rotation. The coefficient of friction between the coin and the turntable is 0.1.

(a) Describe what happens to the coin.

(b) What is the greatest distance from the centre of rotation that the coin can be placed without slipping?

(c) How would your answers to **(b)** and **(c)** change for a heavier coin?

12 A penny, of mass m kg, is placed on the turntable of a record player 0.1 m from the centre. The turntable rotates at 45 rpm. If the penny is on the point of slipping, calculate the coefficient of friction between the penny and the turntable. Calculate the resultant force acting on the penny, when the turntable is rotating at 33 rpm.

13 The position vector **r** (metres), of a particle P at time t s, is given by

$$\mathbf{r} = 2\cos 3t\,\mathbf{i} + 2\sin 3t\,\mathbf{j}$$

where unit vectors **i** and **j** are perpendicular. The mass of the particle is 4 kg.

(a) Show that the particle moves with constant speed.

(b) Find the speed of P.

(c) Find the angular speed of P.

(d) Find the resultant force on P.

14 A car travels without slipping at 10 m s^{-1} around a horizontal bend of radius 30 m. Find the least value of the coefficient of friction.

15 The coefficient of friction between a road surface and the tyres on a car is 0.9. A horizontal bend on the road has a radius of 40 m. Find the maximum speed that the car can take the bend without sliding.

16 A fairground ride consists of a rotating cylinder. People stand on the inside of the cylinder with their backs to it. When the speed of rotation is great enough the floor is lowered so that only friction stops them from falling. The diameter of the cylinder is 10 m. The coefficient of friction between a body and the cylinder is taken as 0.5. Find the least angular speed necessary to stop the body from falling.

17 A child sits on a roundabout at a distance of 5 m from the centre of rotation and at a height of 2 m above ground level. The roundabout completes one revolution every 2 s. After one revolution the child drops a small toy, of mass 500 g, that she was carrying, which then falls to the ground without hitting the roundabout.

(a) Find the acceleration of the child on the roundabout and the magnitude of the force exerted on the toy by the child, before she drops it.

(b) Find the time that it takes for the toy to fall to the ground.

(c) Sketch a graph to show how the magnitude of the acceleration of the toy varies with time. Assume $t = 0$ one revolution before the toy is dropped. [A]

18 A strip of smooth metal, in the shape of a semicircle of radius 20 cm is fixed on a smooth horizontal surface. A marble of mass 20 g is fired into the semicircle and travels at a speed of 5 m s^{-1}. Part of the path of the marble is shown by the dashed line in the diagram.

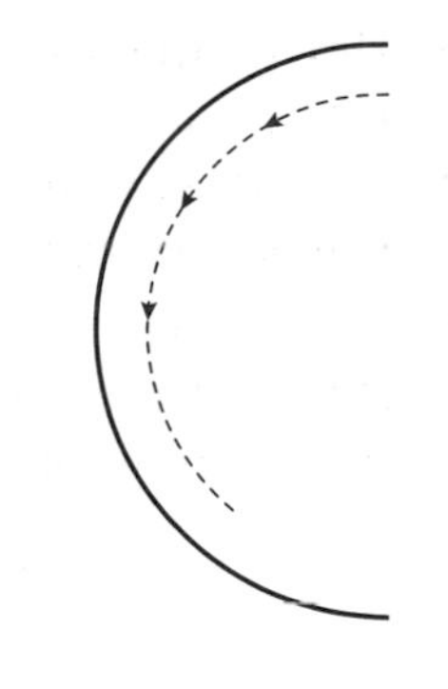

(a) Find the magnitude of the acceleration of the marble in m s^{-2}.

(b) Show that the magnitude of the resultant of the reaction forces acting on the marble is approximately 2.51 N.

(c) Copy the diagram and show the path of the marble when it leaves the semicircle. [A]

19 A device in a funfair consists of a hollow circular cylinder of radius 3 m, with a horizontal floor and vertical sides. A small child stands inside the cylinder and against the vertical side. The cylinder is rotated about its vertical axis of symmetry.

When the cylinder is rotating at a steady angular speed of 30 rev min^{-1} the floor of the cylinder is lowered, so that the child is in contact only with the vertical side.

Given that the child does not slip, find, correct to two decimal places, the minimum coefficient of friction between the child and the side. [A]

20 A traffic roundabout has a radius of 80 m. The road surface at the roundabout is horizontal.

(a) Find the magnitude of the resultant force that acts on a car, of mass 1200 kg, travelling round the roundabout at 20 m s^{-1}.

Assume that the only horizontal force that acts on the car is friction. The coefficient of friction between the tyres and the road is μ.

(b) Find an inequality that μ must satisfy for the car to follow a circular path round the roundabout.

Typical values of μ for road vehicles are between 0.6 and 0.8.

(c) Calculate a safe speed limit for the roundabout.

(d) What factor have you omitted from your calculation that would have reduced the speed limit found in **(c)**? [A]

21 A car of mass 1.2 tonnes is travelling around a roundabout, at a steady speed of 12 m s^{-1}. The friction force that is acting on the car has a magnitude equal to 90% of the magnitude of the normal reaction on the car. Assume that the car can be modelled as a particle and that the road surface is horizontal.

(a) Draw a diagram to show the forces acting on the car if there is assumed to be no air resistance.

(b) Find the radius of the circle described by the car as it travels around the roundabout.

(c) The diagram shows the air resistance force that actually acts on the car as it moves on the roundabout, but that has been ignored in **(a)** and **(b)**.

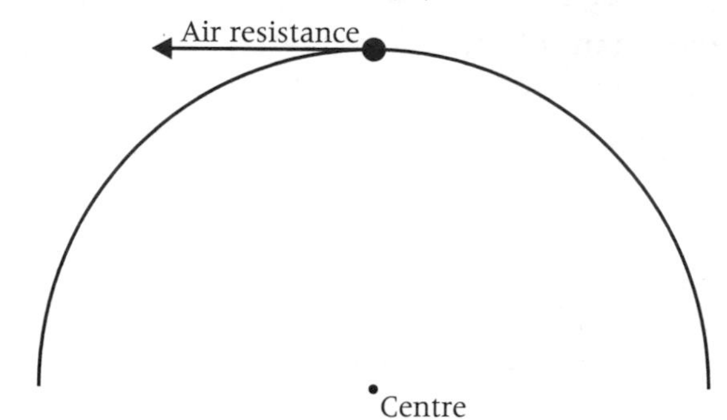

On a copy of the diagram draw a vector to show the resultant force on the car, and hence a vector to show the direction of the friction force on the car (i.e. the force between the tyres and the road). [A]

8.4 Further circular motion

In this section you will consider examples of circular motion in situations where the forces acting are not just horizontal and vertical. The first of these examples is a simple situation, known as the conical pendulum.

Conical pendulum

Consider a particle, of mass m kg, which is suspended from a fixed point A by a light, inextensible string of length l. If the particle moves in a horizontal circle, then the string describes the curved surface of a cone.

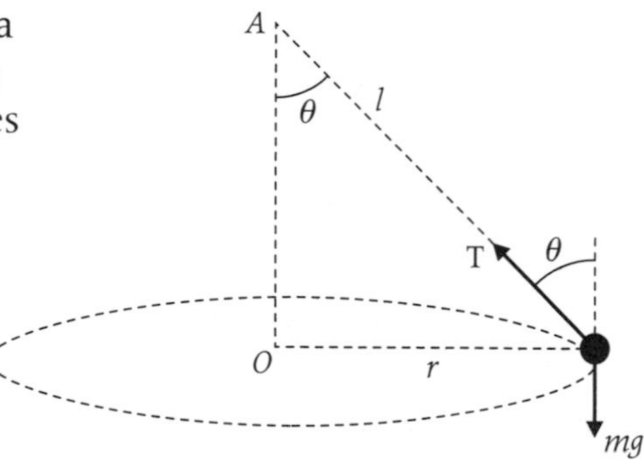

The centre of the circle O will be vertically below A. The radius of the circle will be $l \sin \theta$, where θ is the angle between the string and the vertical.

Using $F = ma$ radially: $T \sin \theta = mr\omega^2$

or $T \sin \theta = ml \sin \theta\, \omega^2$

$T = ml\omega^2$

Resolving vertically: $T \cos \theta = mg$

From these two expressions it is possible to eliminate T to give

$$\cos \theta = \frac{mg}{ml\omega^2} = \frac{g}{l\omega^2}.$$

Note from this result that the angle does not depend on the mass. It also allows us to predict the angle for various combinations of l and ω.

Worked example 8.5

A ball, of mass 400 g, is attached to one end of a light inelastic string of length 0.75 cm. The other end of the string is fixed to a point A. The ball moves in a horizontal circle, at 120 rpm. Find the tension in the string and the angle between the string and the vertical.

Solution

The diagram shows the forces acting on the ball.

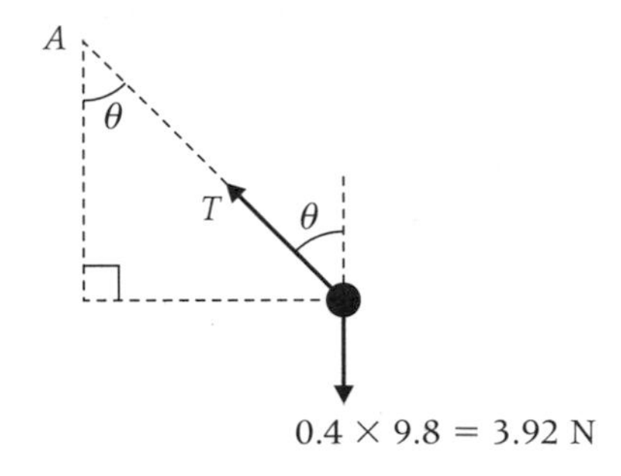

First convert the angular speed to rad s^{-1}.

$$\omega = \frac{120 \times 2\pi}{60} = 4\pi \text{ rad s}^{-1}$$

Using $F = ma$ radially gives

$$T \sin\theta = 0.4 \times 0.75 \sin\theta \times (4\pi)^2$$
$$T = 47.37 \text{ N}$$

Resolving vertically

$$T\cos\theta = 0.4g$$
$$\cos\theta = \frac{0.4 \times 9.8}{47.37}$$
$$\theta = 85.3^\circ$$

Worked example 8.6

A particle P, of mass m, is connected to two light inextensible strings of equal length l. The other ends of the strings are attached to fixed points A and B, a distance l apart with A vertically above B. The particle P rotates in a horizontal circle with angular speed ω, and both strings taut. Find the tension in each string and show that

$$\omega^2 > \frac{2g}{l}$$

Solution

The diagram shows the two tensions that act, along with the weight of the particle.

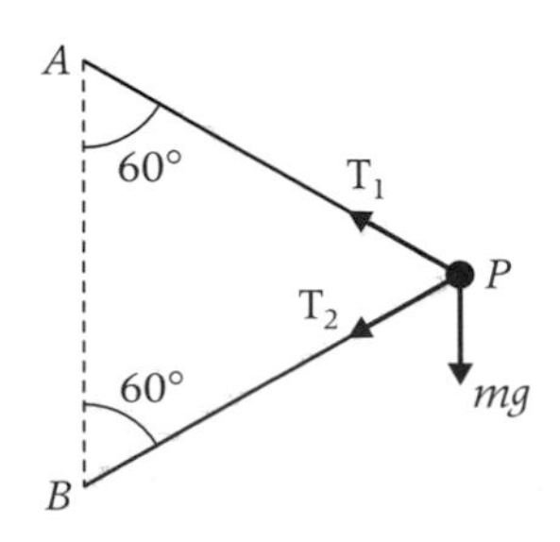

Using $F = ma$ radially gives

$$T_1 \sin 60^\circ + T_2 \sin 60^\circ = ml \sin 60^\circ \omega^2$$
$$T_1 + T_2 = ml\omega^2$$

Resolving vertically gives

$$T_1 \cos 60^\circ = T_2 \cos 60^\circ + mg$$
$$T_1 - T_2 = 2mg.$$

You now have a pair of simultaneous equations which when added give

$$2T_1 = ml\omega^2 + 2mg$$
$$T_1 = mg + \tfrac{1}{2}ml\omega^2$$

Subtracting the equations gives

$$2T_2 = ml\omega^2 - 2mg$$
$$T_2 = \tfrac{1}{2}ml\omega^2 - mg$$

So the tensions in the strings are functions of m, g and ω. In particular the tension in the top string is positive for all values of ω, but this is not the case in the lower string. If the angular speed is not great enough the lower string becomes slack. Hence if both strings are taut

$$T_2 > 0$$
$$\tfrac{1}{2}ml\omega^2 - mg > 0$$
$$\omega^2 > 2\frac{g}{l}$$

Worked example 8.7

A sphere P, of mass m, moves in a horizontal circle,with angular velocity ω, on the inside, smooth surface of an inverted cone, of semi-vertical angle α as shown. Find the radius of the circle in terms of α, ω and g.

Solution

The diagram shows the two forces acting on the sphere, which are modelled as a particle. These forces are the weight and the normal reaction, which is perpendicular to the surface of the cone.

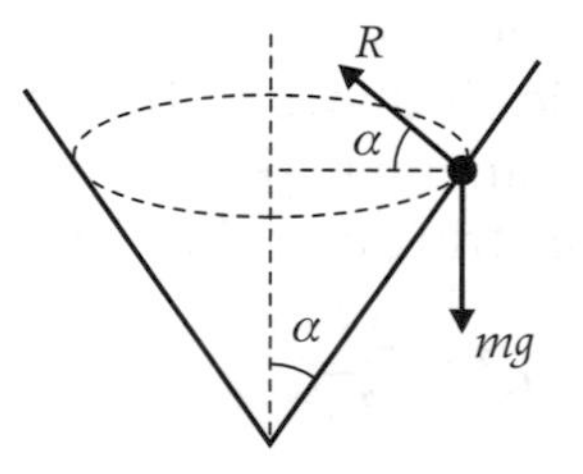

Resolving vertically gives

$$R \sin \alpha = mg$$

Using $F = ma$ radially gives

$$R \cos \alpha = mr\omega^2$$

These two expressions can then be divided as below

$$\frac{R \sin \alpha}{R \cos \alpha} = \frac{mg}{mr\omega^2}$$

$$\tan \alpha = \frac{g}{r\omega^2}$$

Now rearrange to make r the subject of this expression.

$$r = \frac{g}{\omega^2 \tan \alpha}$$

Worked example 8.8

An aircraft banks as it turns in a horizontal circle of radius 500 m. If the speed of the aircraft is 300 km h^{-1}, find the angle to the horizontal at which the aircraft must be banked.

Solution

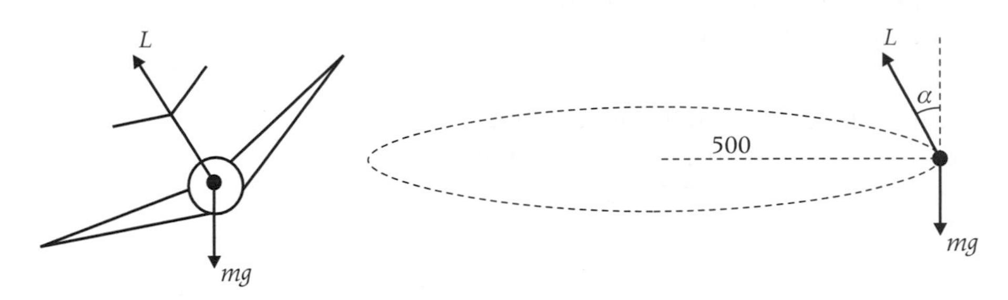

To model this situation assume that the lift force acts through the centre of gravity of the aircraft. The aircraft can thus be treated as a particle. The diagram shows the forces acting on the aeroplane.

Resolving vertically gives

$$L \cos \alpha = mg$$

Using $F = ma$ radially gives

$$L \sin \alpha = m \times \frac{v^2}{500}$$

Where $v = 300 \text{ km h}^{-1} = \frac{300 \times 1000}{3600} \text{ m s}^{-1} = 83.3 \text{ m s}^{-1}$

Dividing both equations and substituting for v gives

$$\frac{\cos \alpha}{\sin \alpha} = \frac{500g}{v^2} = \frac{500 \times 9.8}{83.3^2} = 0.706$$

$\tan \alpha = 1.42$
$\alpha = 55°$ to the nearest degree.

EXERCISE 8C

1 A particle, of mass 2 kg, is attached to a fixed point, A, by a light inextensible string of length l m, which is inclined at an angle θ to the vertical. The particle moves in a horizontal circle with speed v m s^{-1} and angular speed ω. The tension in the string is T N.

(a) If $l = 0.5$ m and $\theta = 20°$, find T and ω.

(b) If $l = 1$ m and $\omega = 5$ rad s^{-1}, find T and θ.

(c) If $v = 2$ m s^{-1} and $\theta = 60°$, find l and T.

2 A particle, of mass 2 kg, is attached to a fixed point, A, by a light inextensible string of length 50 cm. The particle moves in a horizontal circle of radius 10 cm. Find the tension in the string and the particle's angular speed.

3 A particle, of mass 3 kg, is attached to a fixed point, A, by an inelastic string of length 70 cm. The particle moves in a horizontal circle with angular speed of 60 rpm. Find the tension in the string and the radius of the circle.

4 A particle, P, of mass 4 kg, is attached by a light inextensible string, of length 3 m, to a fixed point. The particle moves in a horizontal circle with an angular speed of 2 rad s^{-1}. Calculate:

(a) the tension in the string,

(b) the angle the string makes with the vertical,

(c) the radius of the circle.

5 A particle, of mass 5 kg, is attached to a fixed point by a string of length 1 m. It describes horizontal circles of radius 0.5 m. Calculate the tension in the string and the speed of the particle.

6 A bead, of mass m, is threaded on a string of length 8 m, which is light and inextensible. The free ends of the string are attached to two fixed points separated vertically by a distance, which is half the length of the string, the lower fixed point being on a smooth, horizontal table. The bead is made to describe horizontal circles on the table around the lower fixed point, with the string taut. What is the maximum value of ω, the angular speed of the bead, if it is to remain in contact with the table?

7 A fairground ride consists of a platform, which rotates horizontally. Ropes hang from the platform and people sit in cradles, suspended by the ropes. One rope is 5 m long and hangs from a point 5 m from the centre of rotation. A child sits in the cradle, which rotates in a horizontal circle. If the angle between the rope and the vertical is $\tan^{-1}\left(\frac{3}{4}\right)$ find the angular speed of the ride.

8 Two particles of mass m and $2m$ are connected by a light inextensible string which passes through a smooth fixed ring. The heavier particle hangs in equilibrium below the ring and the lighter particle describes a horizontal circle of radius r. Find the angular speed of the lighter particle in terms of r and g.

9 One end of a light inextensible string of length l is fixed at a point A and a particle P of mass m is attached to the other end. The particle moves in a horizontal circle with constant angular speed ω. Given that the centre of the circle is vertically below A and that the string remains taut with AP inclined at an angle α to the downward vertical, find $\cos \alpha$ in terms of l, g and ω. [A]

10 A particle P is attached to one end of a light inextensible string of length 0.125 m, the other end of the string being attached to a fixed point, O. The particle describes, with constant speed, and with the string taut, a horizontal circle whose centre is vertically below O. Given that the particle describes exactly two complete revolutions per second find, in terms of g and π, the cosine of the angle between OP and the vertical. [A]

11 A particle P of mass m moves in a horizontal circle, with uniform speed v, on the smooth inner surface of a fixed thin, hollow hemisphere of base radius a. The plane of motion of P is a distance $\frac{a}{4}$ below the horizontal plane, containing the rim of the hemisphere. Find, in terms of m, g and a, as appropriate, the speed v and the reaction of the surface on P.

A light inextensible string is now attached to P. The string passes through a small smooth hole at the lowest point of the hemisphere, and has a particle of mass m hanging at rest at the other end. Given that P now moves in a horizontal circle, on the inner surface of the hemisphere with uniform speed u, and that the plane of the motion is now distant $\frac{a}{2}$ below the horizontal plane of the rim, prove that:

$$u^2 = 3ga$$ [A]

12 The point A is vertically above point B, and a distance $5a$ from it. A particle P of mass m is attached to A by a light inextensible string of length $4a$. The particle is also attached to B by a light inextensible string of length $3a$. P moves in a horizontal circle with both strings taut. Find the tension in the strings and show that:

$$\omega^2 \geqslant \frac{5g}{16a}$$ [A]

13 A particle moves with constant speed u in a horizontal circle of radius a on the inside of a fixed smooth spherical shell of internal radius $2a$. Show that $u^2\sqrt{3} = ag$. [A]

14 An aeroplane of mass m kg describes a horizontal circle of radius r m at a constant speed of v m s^{-1}.

(a) Find the magnitude of the resultant force on the aeroplane if $m = 2000$, $v = 50$ and $r = 500$.

(b) A lift force of magnitude L N acts on the aeroplane. This force lies in the vertical plane that contains the aeroplane and the centre of the circle and acts at an angle α to the vertical when the aeroplane is flying in a circle at a constant speed. Find L and α in terms of v, r and g.

(c) Describe how L and α would change if the radius of the circle were reduced, but the speed of the aeroplane remained unchanged. [A]

15 A child of mass 40 kg swings on the end of a rope of length 4 m, moving in a horizontal circle at a constant speed. The rope is at angle of 30° to the vertical. The centre of the circle is at a height of 1.5 m above a point O on the ground. Model the child as a particle at the end of the rope.

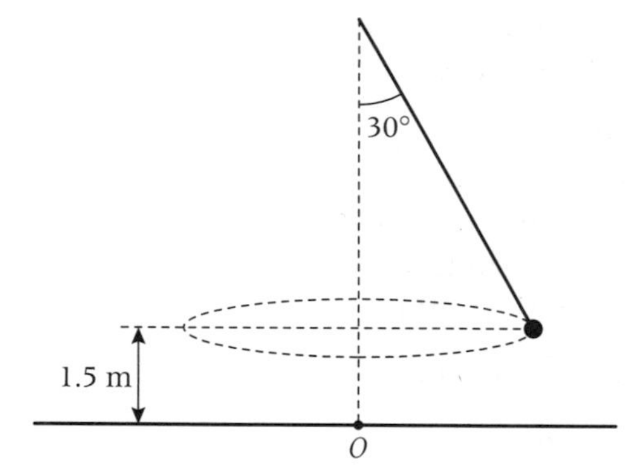

(a) Find

(i) the tension in the rope,

(ii) the acceleration of the child,

(iii) her speed.

The child lets go of the rope and falls freely under gravity.

(b) Find the time that it takes for the child to reach the ground and state a reason why, in reality, this is an overestimate.

(c) Find the maximum distance that the child could land from the point O. Draw a diagram to show the region in which the child could land and state the distance of the boundaries of this region from the point O. [A]

16 A conical pendulum consists of a string of length 0.5 m, with a particle of mass 3 kg attached at one end. The other end of the string is fixed to a point O. The particle describes a horizontal circle at constant speed. The centre of the circle is at the point Q, vertically below O. The string makes an angle of 30° with the vertical as shown in the diagram. Take $g = 10\text{ m s}^{-2}$.

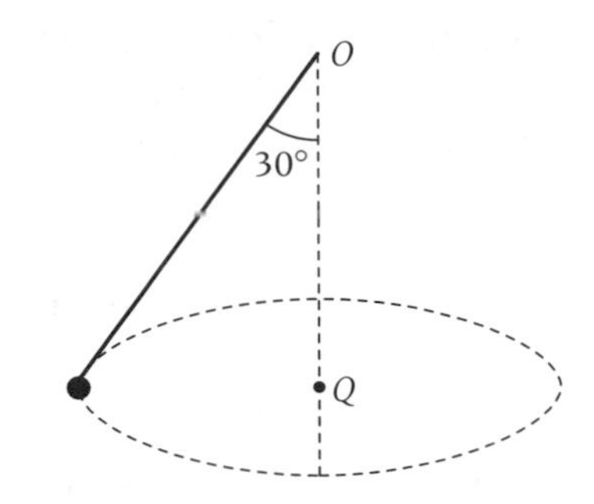

(a) State **two** assumptions that should be made about the string in order to form a simple model of the situation.

(b) (i) Show that the tension in the string is $20\sqrt{3}$ N.

(ii) If ω radians per second is the angular speed of the particle, show that $\omega^2 = \dfrac{40\sqrt{3}}{3}$.

A second string, of the same length, is attached to the particle and to a point P, vertically below Q, such that $OQ = QP$, as shown in the diagram. The particle describes the same circle as in **(b)**, but at twice the angular speed.

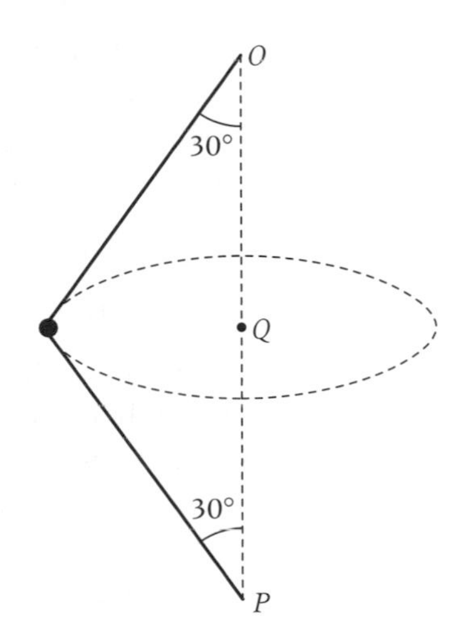

(c) Find the tension in each string. [A]

17 A particle, of mass m, is attached by a light, inextensible string, of length l, to the top of a smooth cone. The particle is set into motion so that it describes a horizontal circle on the outer surface of the cone as shown in the diagram below.

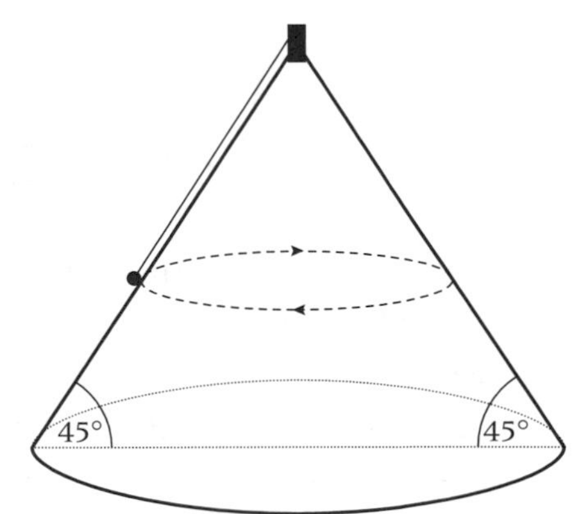

(a) Show that the tension in the string has magnitude $m\left(\dfrac{v^2}{l} + \dfrac{g}{\sqrt{2}}\right)$, when the particle describes a circle at a constant speed v.

(b) Find the magnitude of the normal reaction force that the cone exerts on the particle.

(c) What will happen if $v^2 > \dfrac{gl}{\sqrt{2}}$? Justify your answer. [A]

18 A fairground ride consists of a circular drum. The drum is a circular cylinder with vertical sides and of radius 2 m. It rotates about a vertical axis, and when it has reached its normal operating speed, the floor of the drum is removed so that the people on the ride move in a horizontal circle 'stuck' to the inside of the drum. At its normal operating speed, the drum completes 1 revolution every 2 s. Model a person on the ride as a particle.

(a) Find the acceleration of a person on the ride, at its normal operating speed.

(b) Draw and label a diagram to show the forces acting on a person on the ride, when the floor has been removed. Find the magnitude of the reaction and friction forces acting on a person of mass 70 kg in this situation.

(c) The operators propose a new design in which the sides of the drum are at angle θ to the horizontal as shown in the diagram below.

Assume that people on the ride still move in a circle of radius 2 m and at the same angular speed as before. Show that the magnitude of the friction force, F, acting on a person of mass m, is given by

$$F = mg \sin \theta - 2\pi^2 m \cos \theta.$$

Hence show that F is zero when $\tan \theta = \frac{2\pi^2}{g}$. [A]

19 The diagram shows a truncated cone. The radius of the cone decreases from 40 cm to 30 cm and the height of the truncated cone is 50 cm.

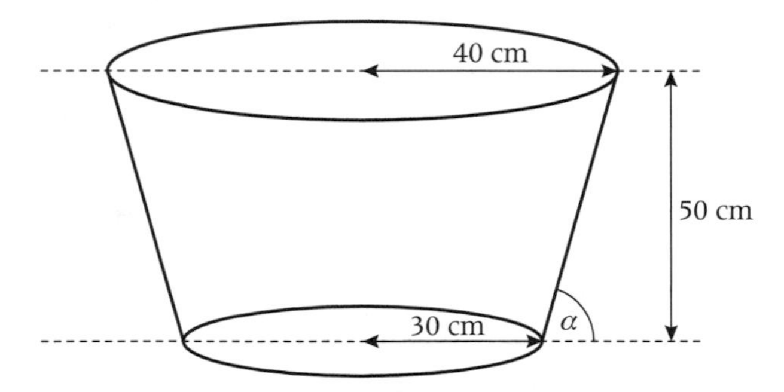

The angle between the sloping surface of the cone and the horizontal is α.

(a) Find $\tan \alpha$.

A coin is projected into the truncated cone, so that it rolls round on the inside surface.

When the coin is inside the truncated cone, it is modelled as a particle that **slides** on a smooth surface and it is assumed that there is no air resistance. The coin is assumed to move in a horizontal circle, on the inside surface of the coin, at a constant speed.

(b) Show that if the coin moves in a circle of radius r m, its speed, v m s^{-1}, is given by $v = \sqrt{rg \tan \alpha}$.

(c) Find the range of speeds for which the coin can be modelled as a particle describing a horizontal circle inside the truncated cone.

In reality the coin slows down, gradually moves down the cone, and eventually drops out of the bottom. Assume that the coin moves in a horizontal circle, of radius 40 cm, when it enters the truncated cone and that later it moves in a circle, of radius 30 cm, just before it leaves the cone. The mass of the coin is 10 g.

(d) Find the total energy lost by the coin, as it moves from the top to the bottom of the truncated cone.

(e) State one aspect of the motion of the coin that has been ignored by the use of the particle model. [A]

8.5 Motion of a vehicle on a banked track

In an earlier section the motion of a car as it turned a circle on a horizontal road was considered. You found that when a car turns such a corner, friction acts towards the centre of the circle and prevents sliding. Friction, of course, has a maximum value, so there is a maximum speed that the car can take the bend without sliding. If the road is banked, however, this maximum speed will increase because the normal reaction will also have a component towards the centre. To model this motion you will consider a particle on a rough inclined plane. It is possible that the frictional force, F, can act either up or down the plane. There are three important cases which need consideration.

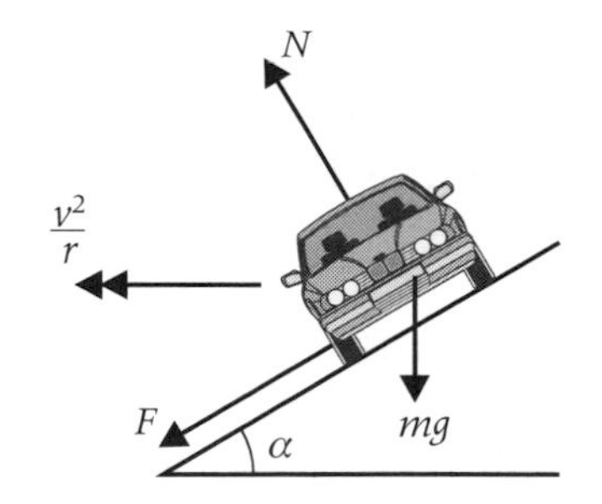

(a) If the car were on the point of sliding up the incline, F, would act down the plane. Consider this when investigating the maximum speed that the car can take the bend.

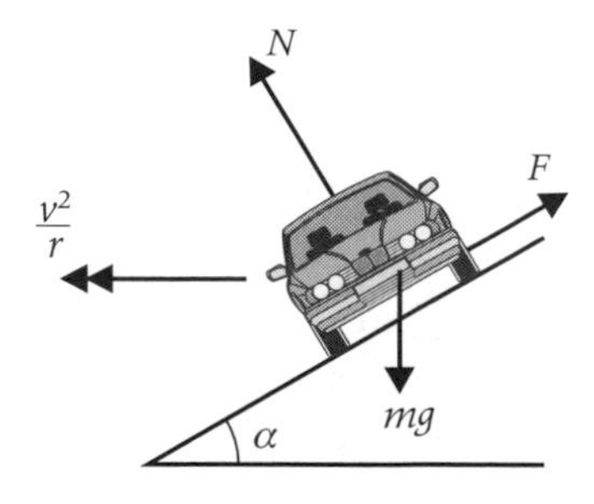

(b) If the car were on the point of sliding down the plane, F, would act up the plane. Consider this when investigating the minimum speed that the car can take the bend.

(c) Between the above two extremes, F can be zero. This is important for various reasons; reduced wear on tyres, greater comfort for passengers, and safety. There will be an optimum speed, which will ensure this.

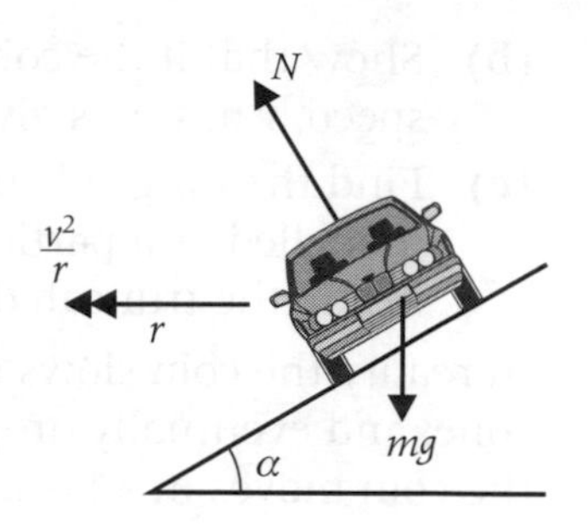

Worked example 8.9

A car takes a bend on a racetrack, which is banked at 10° to the horizontal. Thc radius of the curve is 100 m, and the coefficient of friction between the tyres and road is 0.8. Find the greatest speed that the car can take the bend, if it travels in a horizontal circle.

Solution

The diagram shows the forces acting on the car.

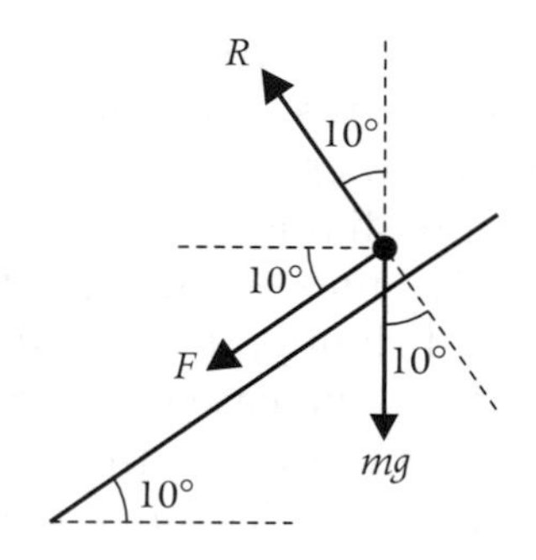

Using $F = ma$ radially gives

$$R \sin 10° + F \cos 10° = \frac{mv^2}{100} \qquad (1)$$

Resolving vertically gives

$$R \cos 10° = mg + F \sin 10° \qquad (2)$$

$$R \cos 10° - F \sin 10° = mg$$

When the speed is a maximum the car will be on the point of sliding, so friction will be limiting, hence $F = \mu R$. This can be used to substitute for F in the two equations above, which then become

$$R \sin 10° + 0.8\, R \cos 10° = \frac{mv^2}{100}$$

$$R \cos 10° - 0.8\, R \sin 10° = mg$$

If we now divide these two equations, then R will cancel on the left-hand side.

$$\frac{R \sin 10° + 0.8\, R \cos 10°}{R \cos 10° - 0.8\, R \sin 10°} = \frac{mv^2}{100mg}$$

$$\frac{\sin 10° + 0.8 \cos 10°}{\cos 10° - 0.8 \sin 10°} = \frac{v^2}{100g}$$

$$v^2 = 100g\left(\frac{\sin 10° + 0.8 \cos 10°}{\cos 10° - 0.8 \sin 10°}\right)$$

$$v = 33.4 \text{ m s}^{-1} \text{ (to 3 sf)}$$

Note: equations (1) and (2) are a pair of simultaneous equations for R and F, which can be solved to get the values of R and F.

(1) × sin 10° + (2) × cos 10° gives

$$R(\sin^2 10° + \cos^2 10°) = \frac{mv^2 \sin 10°}{100} + mg \cos 10°$$

$$R = \frac{mv^2 \sin 10°}{100} + mg \cos 10°$$

Similarly (1) × cos 10° − (2) × sin 10° gives

$$F(\sin^2 10° + \cos^2 10°) = \frac{mv^2 \cos 10°}{100} - mg \sin 10°$$

$$F = \frac{mv^2 \cos 10°}{100} - mg \sin 10°$$

Worked example 8.10

The radius of a bend in a road is r, and the road is banked at α to the horizontal. Find the speed at which a vehicle should take the bend, so that there is no tendency to slip.

Solution

The diagram shows the forces acting on the vehicle when it is modelled as a particle. Note that because there is no tendency to slip a friction force has not been included.

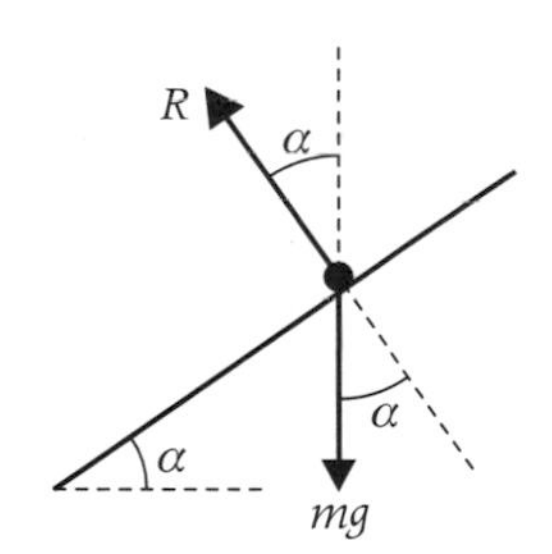

Using $F = ma$ radially gives

$$R \sin \alpha = \frac{mv^2}{r}$$

Resolving vertically

$$R \cos \alpha = mg$$

Dividing these two equations gives

$$\frac{\sin \alpha}{\cos \alpha} = \tan \alpha = \frac{v^2}{rg}$$

So, the optimum speed is given by

$$v^2 = rg \tan \alpha.$$

The vehicle in the previous example was not specified. The optimum speed found is particularly applicable to locomotives. When a train turns a corner on a piece of horizontal track, friction does not provide the horizontal force, but the rails do. This force can be very large, because of the large mass of the train. If a train needs to turn at high speeds, then it is necessary to bank the track, for safety.

Worked example 8.11

A vehicle, of mass 1 tonne, takes a bend of radius 50 m, on a horizontal road without slipping. The car travels at 25 m s^{-1}, and the road is banked at 15° to the horizontal. The car travels in a horizontal circle. Find the frictional force and the normal reaction of the road on the car. What is the minimum value of the coefficient of friction?

Solution

The diagram shows the forces acting on the vehicle, when it is modelled as a particle.

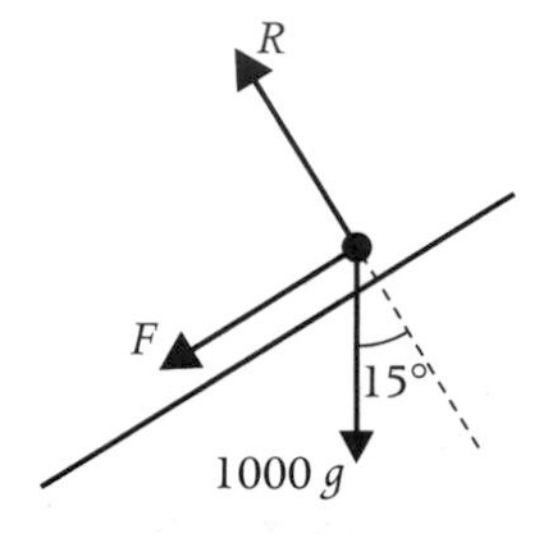

The first consideration is that we do not know whether the frictional force acts up or down the plane. However, this does not cause a problem. We can assume that it acts down the plane, and if we have chosen incorrectly our answer will turn out negative, but will still have the correct magnitude.

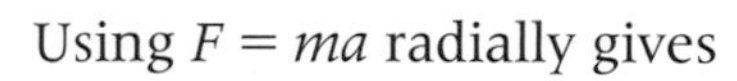

Using $F = ma$ radially gives

$$R \sin 15° + F \cos 15° = 1000 \times \frac{25^2}{50}$$

$$R \sin 15° + F \cos 15° = 12\,500 \qquad (1)$$

Resolving vertically gives

$$R \cos 15° = 1000\,g + F \sin 15°$$

$$R \cos 15° - F \sin 15° = 9800 \qquad (2)$$

To find R we calculate $(1) \times \sin 15° + (2) \times \cos 15°$ gives

$$R\,(\sin^2 15° + \cos^2 15°) = 12\,500 \sin 15° + 9800 \cos 15°$$

$$R = 12.7\text{ kN}$$

Similarly $(1) \times \cos 15° - (2) \times \sin 15°$ gives

$$F\,(\sin^2 15° + \cos^2 15°) = 12\,500 \cos 15° - 9800 \sin 15°$$

$$F = 9.54\text{ kN}$$

Using $F \leqslant \mu R$ gives

$$9.54 \leqslant 12.7\mu$$

$$\mu \geqslant 0.751$$

EXERCISE 8D

1 A car is negotiating a bend of radius 100 m, banked at an angle of 5°. What is the maximum speed at which it can do this, if the coefficient of friction is 1.0, the car does not slip, and the path of the car is in a horizontal plane?
At what angle must the bend be banked if the car is to negotiate the bend without any tendency to slip at a speed of 80 km h^{-1}?

2 A stunt man is required to drive a car round a tightly banked corner, in a horizontal circle. The track is banked at 60° to the horizontal, and the radius of the bend is 30 m, and the coefficient of friction is 0.5. Find the minimum speed that he must drive if there is to be no slipping.

3 A train of mass 100 tonnes travels round a banked track, which forms part of a circle of radius 1250 m. The distance between the rail centres is 1.5 m and the outer rail is 37.5 mm vertically higher than the inner rail. If the train is travelling at 63 km h^{-1}, show that the sideways force on the rails is zero.

4 One lap of a circular cycle track is 400 m, and the track is banked at 45°. At what speed can the track be negotiated without any tendency to slip?

5 The radius of a bend on a horizontal piece of railway track is 750 m. The distance between the centres of the rails is 1.5 m. The average speed of trains on this stretch of track is 120 km h^{-1}. Find the height of the outer rail above the inner rail, if a train travelling at the average speed is to exert no sideways force on the rails.

6 A vehicle of mass 950 kg takes a bend of radius 60 m, without slipping. The speed of the vehicle is 20 m s^{-1}, and the road is banked at 20° to the horizontal. The vehicle travels in a horizontal circle. Find the frictional force and the normal reaction of the road on the vehicle. What is the least value of the coefficient of friction?

7 A car undergoing trials is moving on a horizontal surface around a circular bend of radius 50 m at a steady speed of 14 m s^{-1}. Calculate the least value of the coefficient of friction between the tyres of the car and the road surface.
Find the angle to the horizontal at which the bend should be banked in order that the car can move in a horizontal circle of radius 50 m at 14 m s^{-1} without any tendency to slip.
Another section of the test area is circular and banked at 30° to the horizontal. The coefficient of friction between the tyres of the car and the road surface is 0.6. Calculate the greatest speed at which the car can move in a horizontal of radius 70 m around the banked test area. [A]

8 A car, of mass 1200 kg, travels round a bend, of radius 50 m, at a constant speed of 20 m s^{-1}. Model the car as a particle and assume that there is no air resistance acting on it.

(a) A simple model assumes that the road is horizontal. Find the magnitude of the friction force acting on the car.

(b) In reality the road is banked at 2° to the horizontal. Find the magnitude of the friction force acting on the car.

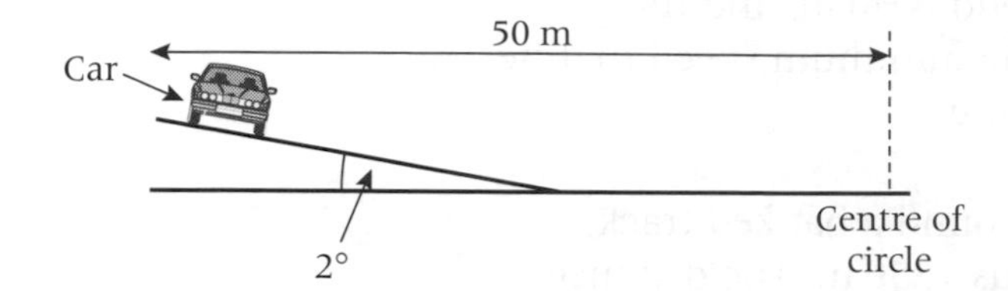

Draw and label a diagram to show the forces on the car and find a revised value for the magnitude of the friction force that takes account of the banking. [A]

8.6 Motions in a vertical circle at constant speed

In this chapter a simple case of vertical circular motion, in which the speed of the object moving in the circle remains constant will be considered. An example of this would be an item of clothing that moves in a vertical circle during the drying phase of a washing machine cycle.

Worked example 8.12

A jumper has a mass of 0.4 kg when wet. It is inside a washing machine drum that has a radius of 35 cm and rotates at 1200 rpm. Model the jumper as a particle that travels in a vertical circle. Find the magnitude of the reaction force on the jumper at its highest and lowest positions.

Solution

First we will convert the angular speed from rpm to rad s^{-1}.

$$\omega = \frac{1200 \times 2\pi}{60}$$
$$= 40\pi \text{ rad s}^{-1}$$

The diagram shows the forces acting on the jumper at its highest point. As both these forces act towards the centre of the circle we can apply $F = ma$ radially to give

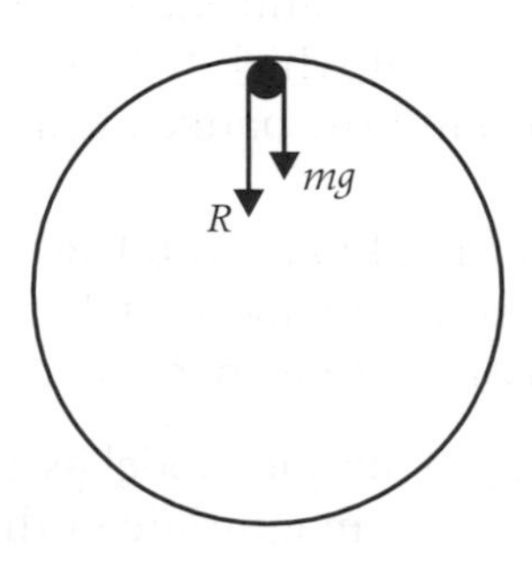

$$R + 0.4 \times 9.8 = 0.4 \times 0.35 \times (40\pi)^2$$
$$R = 2207 \text{ N}$$

The second diagram shows the forces acting at the lowest point. In this case we can apply $F = ma$ radially to give

$$R - 0.4 \times 9.8 = 0.4 \times 0.35 \times (40\pi)^2$$
$$R = 2214 \text{ N}$$

We can see that the reaction force only varies by a very small amount between these two positions.

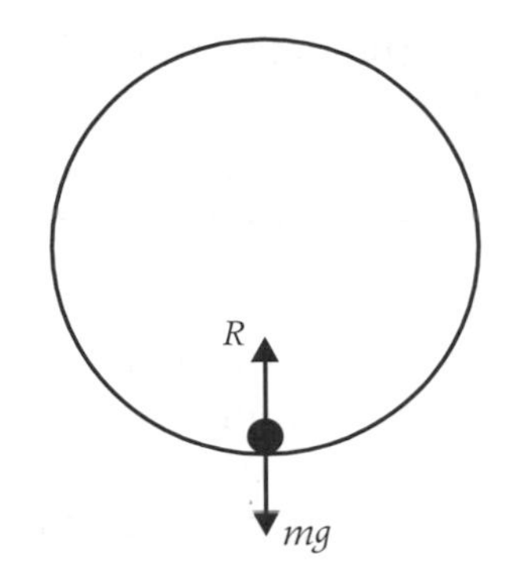

Worked example 8.13

A drum has a radius of 2 m. Initially the drum rotates around a vertical axis with people standing against the inside surface of the drum. When the drum reaches a certain speed it is moved so that it rotates around a horizontal axis with the people in it travelling in vertical circles. Find the minimum safe angular speed for the drum.

Solution

A force of magnitude $mr\omega^2$, must always act towards the centre of the circle. This force will be provided by the reaction force from the surface of the drum and the component of gravity acting towards the centre of the circle.

The diagram shows the forces acting in one position.

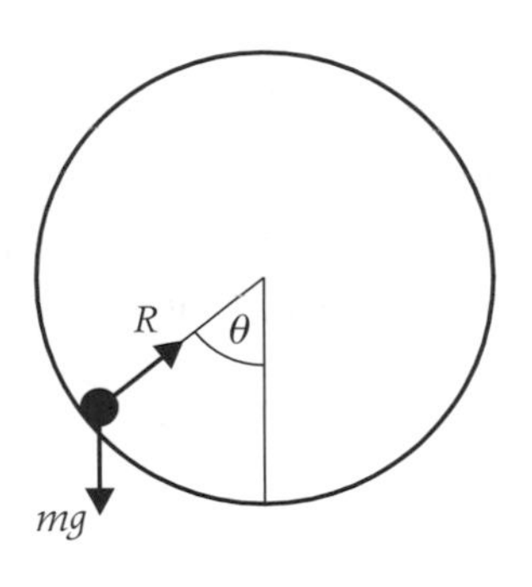

Apply $F = ma$ radially gives

$$R - mg \cos \theta = mr\omega^2$$
$$R = mr\omega^2 + mg \cos \theta$$

The person in the drum will complete vertical circle provided they remain in contact with the inside of the drum, that is if $R \geqslant 0$ for all values of θ.

$$mr\omega^2 + mg \cos \theta \geqslant 0$$
$$mr\omega^2 \geqslant - mg \cos \theta$$

As this is true for all values of θ, we must take the minimum value of $\cos \theta$, which is -1 to give

$$mr\omega^2 \geqslant mg$$

$$\omega \geqslant \sqrt{\frac{g}{r}}$$

Substituting $r = 2$ gives

$$\omega \geqslant \sqrt{4.9} = 2.21 \text{ rad s}^{-1}$$

EXERCISE 8E

1 An item of clothing inside a washing machine is modelled as a particle, of mass 0.6 kg. The drum of the washing machine has radius 60 cm and rotates at 900 rpm. Assume that the clothing travels in a vertical circle at a constant speed.

8

Find the magnitude of the reaction force on the clothing, when it is

(a) at its highest point,
(b) level with the centre of the drum,
(c) at its lowest point.

2 In a film a stuntman, of mass 80 kg, holds onto the end of the sail of a windmill, so that he describes a vertical circle of radius 4 m, moving at a constant speed. Model the stuntman as a particle at the end of one of the sails. The sails complete one revolution every 10 s.

Find the magnitude of the force that the stuntman must exert at his lowest point.

3 A disc of radius 2.5 m rotates at 120 rpm about a horizontal axis. A small object, of mass 5 kg, is attached to the edge of the disc with Velcro, so that it describes a vertical circle.

(a) Find the maximum force that the Velcro must exert on the disc if the object is to travel in a circle at this speed.

(b) If the Velcro can only provide a force of magnitude 2000 N, describe the position of the object when it leaves contact with the disc.

4 An aeroplane of mass 3000 kg, that loops the loop is to be modelled as a particle that describes a circle of radius 200 m while travelling at a constant 60 m s^{-1}.

Assume that due to the wings a force, L N, acts on the plane and is always directed towards the centre of the circle.

Find L, when the plane is at each of the positions shown in the diagram, where OA is horizontal.

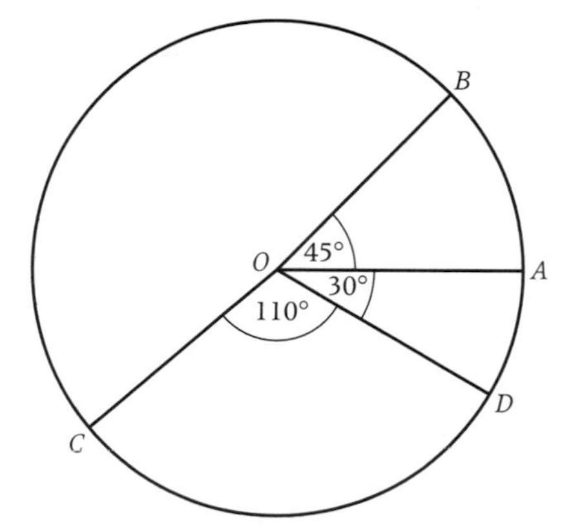

5 The wheel of a car has radius 30 cm. At the centre of the wheel is a hub cap of radius 15 cm. A stone is placed inside the hub cap, so that when the car goes fast enough the stone describes a vertical circle in contact with the inside surface of the hub cap. Find the minimum speed of the car for which the stone will remain in contact with the surface of the hub cap.

Key point summary

1 $1 \text{ rpm} = \frac{2\pi}{60} \text{ rad s}^{-1}$. p146

2 The velocity has magnitude $r\omega$ and is directed along a tangent to the circle. p146

3 It can be useful to express the magnitude of acceleration in terms of speed, v, rather than ω. p147

From $v = r\omega$, $\omega = \frac{v}{r}$ and by substituting this, the expression for the magnitude of acceleration becomes

$$a = r\left(\frac{v}{r}\right)^2 = \frac{v^2}{r}.$$

4 When a particle moves in a horizontal circle with constant speed, there are two principles that can always be applied: p149

- resultant of vertical components of forces must be zero
- $F = ma$ can be applied radially.

8

Test yourself

What to review

1 A coin, of mass 20 g, is placed on a turntable that rotates at 50 rpm. The coefficient of friction between the coin and the turntable is 0.6.

(a) Find the maximum distance that the coin can be placed from the centre of the turntable if the coin is to travel in a circle.

(b) Find the magnitude of the friction force in this case.

Section 8.1

---

2 A ball, of mass 300 g, is attached to one end of a light string of length 60 cm. The ball describes a horizontal circle with the string at an angle of 20° to the vertical.

(a) Find the speed of the ball.

(b) Find the tension in the string.

Section 8.2

---

3 A car travels in a horizontal circle of radius 75 m on a track that is banked at 5° to the horizontal. The coefficient of friction between the car and the track is 0.8.

(a) Find the speed at which the car has no tendency to slip.

(b) Find the maximum speed of the car on the bend.

Section 8.3

---

4 A small sphere, of mass 3 kg, is attached to the end of a light rod of length 1.5 m. The rod rotates at 90 rpm. Find the tension in the rod when it is:

(a) horizontal,

(b) vertical, with the sphere above the centre of the circle,

(c) at an angle of 45° to the vertical with the sphere higher than the centre of the circle.

Section 8.4

Test yourself ANSWERS

1 (a) 21.4 cm, **(b)** 0.118 N.

2 (a) 0.856 m s^{-1}, **(b)** 3.13 N.

3 (a) 8.02 m s^{-1}, **(b)** 26.5 m s^{-1}.

4 (a) 400 N, **(b)** 370 N, **(c)** 379 N.

Exam style practice paper

Time allowed 1 hour 45 minutes

Maximum marks: 80

Answer **all** questions

1 A rocket is fired so that it travels vertically upwards. It is initially at rest and reaches a height of 200 m after 20 s. Assume that the rocket moves with constant acceleration.

(a) Find the acceleration of the rocket. *(3 marks)*

(b) Find the speed of the rocket after 20 s. *(2 marks)*

The mass of the rocket is 250 g.

(c) Find the magnitude of the upward force that acts on the rocket as it moves. *(4 marks)*

2 A boat is initially at point A moving west at 6 m s^{-1}. Two minutes later it is at point B moving due north at 4 m s^{-1}. The unit vectors $\mathbf{i}$ and $\mathbf{j}$ are directed east and north respectively. Assume that the boat moves with constant acceleration between A and B.

(a) Find the acceleration of the boat in terms of $\mathbf{i}$ and $\mathbf{j}$. *(3 marks)*

(b) Find the distance between A and B. *(4 marks)*

(c) What can you deduce about the distance travelled by the boat in the two minute period? *(1 mark)*

3 A driver hangs a toy monkey from the mirror of his car. Model the monkey as a particle that hangs on the end of a light string. As the car moves along a straight road it accelerates uniformly from rest to a speed of 20 m s^{-1} as it travels 80 m.

(a) Calculate the acceleration of the car. *(3 marks)*

(b) Draw a diagram to show the forces acting on the monkey when the car is accelerating. *(1 mark)*

(c) Find the angle between the string and the vertical when the car is accelerating. *(4 marks)*

(d) How would the effects of air resistance on the car alter this angle? *(1 mark)*

Later, when the car is braking the angle between the string and the vertical is 3°.

(e) Find the acceleration of the car when it is braking. *(3 marks)*

4 An aeroplane has position vector $\mathbf{r}$ m relative the point O at time t s.

$$\mathbf{r} = (200 + 50t)\mathbf{i} + (30 - 0.5t^2)\mathbf{j} + 0.8t^2\mathbf{k}$$

The unit vectors **i**, **j** and **k** are directed east, north and vertically upwards, respectively.

(a) Find the distance of the aeroplane from O when $t = 10$. *(2 marks)*

(b) Find an expression for the velocity of the aeroplane at time t. *(2 marks)*

(c) Find the time when the aeroplane is rising at 20 m s^{-1}. *(2 marks)*

(d) Find the time when the aeroplane is travelling south east. *(2 marks)*

5 In a bungee running contest an elastic rope has one end fixed at O. A competitor, of mass 60 kg, is attached to the other end of the rope. The competitor runs to point A, 10 m from O, where he is brought to rest by the rope. The rope has natural length 5 m and modulus of elasticity 400 N.

(a) Calculate the elastic potential energy in the rope when the competitor is at A. *(2 marks)*

At A the competitor falls over and is dragged back by the elastic rope. The coefficient of friction between the competitor and the ground is 0.3.

(b) Find the speed of the competitor when the rope becomes slack. *(3 marks)*

(c) Find the distance of the competitor from O when he comes to rest. *(3 marks)*

6 A lorry has a mass of 40 tonnes and a power output of 400 kW. It ascends a slope inclined at 5° to the horizontal at a maximum speed of 10 m s^{-1}.

(a) Show that the magnitude of the resistance force acting on the lorry when it travels at 10 m s^{-1} is approximately 5800 N. *(3 marks)*

Two possible models for the magnitude of the resistance are to be considered.
The first assumes that the resistance forces are proportional to the speed of the lorry.
The second assumes that the resistance forces are proportional to the square of the speed of the lorry.

(b) For each model predict the maximum speed of the lorry on a horizontal road. *(4 marks)*

(c) State which model is most realistic. Explain why. *(2 marks)*

7 A particle of mass m is attached to two strings. The other ends of the strings are attached at the points A and B. B is directly above A. The particle describes a horizontal circle, of radius r, with A at its centre. The angle between the two strings is 45°.

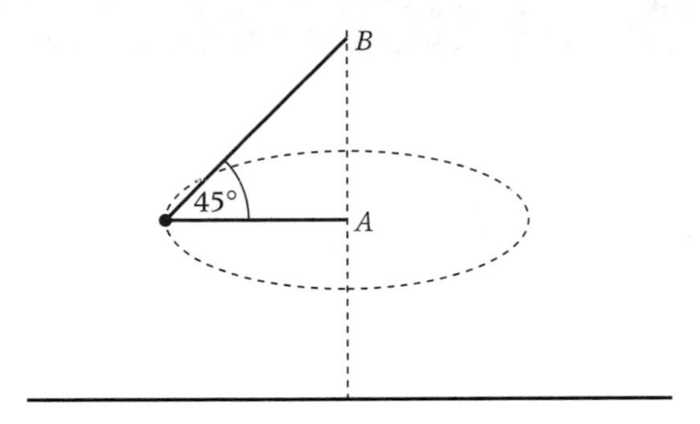

(a) Find the tension in the upper string in terms of m and g. *(4 marks)*

(b) Show that the tension in the lower string is $m\left(\frac{v^2}{r} - g\right)$ and state the range of values of v for which this expression is true. *(4 marks)*

(c) Find the speed of the particle, in terms of r and g, when the tensions in the two strings are equal. *(3 marks)*

When the tensions in the two strings are equal they are released, so that the particle is free to move under gravity only. Point A is a distance $\sqrt{2}r$ above ground level.

(d) Show that the speed of the particle when it hits the ground is $rg\sqrt{1 + 3\sqrt{2}}$. *(4 marks)*.

8 The displacement, x m, of a particle at time t s is

$$x = 5e^{-0.2t}\sin(20t).$$

(a) Find an expression for the velocity of the particle at time t s. *(3 marks)*

(b) Show that the velocity of the particle is zero for the first time when $t = \frac{1}{20}\tan^{-1}100$. *(4 marks)*

(c) Find the speed of the particle when it returns to its initial position for the first time. *(4 marks)*

Answers

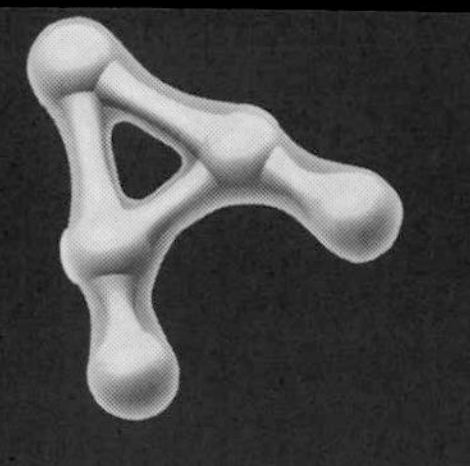

2 Kinematics in one dimension with constant acceleration

EXERCISE 2A

1 84 m, 1.6 m s^{-2}, 0 m s^{-2}, -2 m s^{-2}.

2 104 m.

5 **(a)** 0 m s^{-2}, **(b)** 2 m s^{-2}, **(c)** $-\frac{5}{3}$ m s^{-2}, **(d)** 437.5 m.

6 584 m.

7 0.25 m s^{-2}, 0 m s^{-2}, -0.5 m s^{-2}, 8000 m.

8 **(a)** 16 000 m, **(b)** $233\frac{1}{3}$ s.

9 1.2 m s^{-2}, -1.8 m s^{-2}, 85 s, 1305 m.

10 9200 m, 440 s.

11 **(a)** 10 m s^{-1}, 2.5 m s^{-2}, **(b)** $\frac{16}{9}$ s, 5.06 m s^{-2}.

12 **(a)** 40 m, 20 m s^{-1}, **(b)** 18 s, 6.67 m s^{-2}.

EXERCISE 2B

1 **(a)** 100 m, 20 m s^{-1}, **(b)** 22.5 m s^{-1}, 206.25 m.

2 **(a)** 0.3 m s^{-2}, **(b)** 33.3 s.

3 **(a)** 0.7 m s^{-2}, **(b)** 85 m.

4 **(a)** 0.2 m s^{-2}, **(b)** 10 m, **(c)** 15 m.

5 0.417 m s^{-2}.

6 361 m.

7 -10 m s^{-2}, 1.69 s.

8 **(a)** 889 m, **(b)** 66.7 s.

9 1200 m.

10 **(a)** 2.92 m s^{-2}, 18.7 m s^{-1}, **(b)** 8.55 s.

11 **(a)** 16.3 m s^{-1}, **(b)** 2.21 m s^{-2}.

12 0.580 m s^{-2}, 15.3 m s^{-1}.

13 9.33 s.

14 12 400 m.

15 **(a)** 24 m s^{-1}. **(b)** 7800 m.

16 8 m, 7.5 s.

17 **(a)** 25 m, **(b)** 20 s, **(c)** 50 m s^{-1}/62.5 m s^{-1}.

18 **(a)** $t = 4$ s, **(b)** no, as vehicles are 36 m apart.

19 **(a)** 36 m, **(b)** 36 m + car length.

EXERCISE 2C

1 **(a)** 0.639 s, **(b)** 1.58 s.

2 **(a)** 0.782 s, **(b)** 7.67 m s^{-1}.

3 **(a)** 1.5 s, **(b)** 12.0 m, **(c)** 15.4 m s^{-1}.

4 **(a)** 2 m s^{-2}, 20 m s^{-1}, **(b)** 120.4 m.

6 **(a)** 2.02 s, **(b)** 2.02 s.

7 2.10 s, 24.6 m s^{-1}.

8 6.64 m s^{-1}, -40.1 m s^{-1}.

9 **(a)** The one thrown down, **(b)** both have the same speed, **(c)** 5 m, 5.41 m s^{-1}.

10 32.75 m s^{-1}.

11 $\frac{h}{4}$.

12 **(a)** 12 m, constant acceleration, **(b)** 31.1 m.

13 **(a)** 0.569 m, **(b)** overestimate due to air resistance.

3 Kinematics in two and three dimensions with constant acceleration

EXERCISE 3A

1 **(a)** $0\mathbf{i} + 0\mathbf{j}$, $30\mathbf{i} + 20.1\mathbf{j}$, $60\mathbf{i} + 30.4\mathbf{j}$, $90\mathbf{i} + 30.9\mathbf{j}$, $120\mathbf{i} + 21.6\mathbf{j}$, $150\mathbf{i} + 2.5\mathbf{j}$, $180\mathbf{i} - 26.4\mathbf{j}$, **(c)** 153 m.

2 **(a)** $0\mathbf{i} + 1\mathbf{j}$, $1\mathbf{i} + 1.6\mathbf{j}$, $2\mathbf{i} + 1.8\mathbf{j}$, $3\mathbf{i} + 1.6\mathbf{j}$, $5\mathbf{i} + 0\mathbf{j}$.

3 Children collide.

4 **(a)** $6\mathbf{i} - 9\mathbf{j}$, $86\mathbf{i} + 111\mathbf{j}$, $166\mathbf{i} + 231\mathbf{j}$, $246\mathbf{i} + 351\mathbf{j}$, **(c)** 11.1 m, **(d)** 90 s, 96.7 s.

5 **(a)** 1.225 m, **(b)** 0.5 s, **(c)** 90 m.

6 $16\mathbf{i} + 8\mathbf{j} + 0\mathbf{k}$.

EXERCISE 3B

1 **(a)** $45\cos 10°\mathbf{i} + 45\sin 10°\mathbf{j} = 44.3\mathbf{i} + 44.3\mathbf{j}$,
(b) $105\cos 30°\mathbf{i} + 105\sin 30°\mathbf{j} = 90.9\mathbf{i} + 52.5\mathbf{j}$,
(c) $-21\cos 70°\mathbf{i} + 21\sin 70°\mathbf{j} = -7.18\mathbf{i} + 19.7\mathbf{j}$,
(d) $-62\cos 10°\mathbf{i} - 62\sin 10°\mathbf{j} = -61.1\mathbf{i} - 10.8\mathbf{j}$,
(e) $290\cos 72°\mathbf{i} - 290\sin 72°\mathbf{j} = 89.6\mathbf{i} - 4.24\mathbf{j}$.

2 $-6\cos 45°\mathbf{i} - 6\sin 45°\mathbf{j} = -4.24\mathbf{i} - 4.24\mathbf{j}$.

3 **(a)** $5\mathbf{i}$, **(b)** $-5\mathbf{j}$, **(c)** $-5\mathbf{i}$, **(d)** $5\cos 45°\mathbf{i} - 5\sin 45°\mathbf{j}$,
(e) $5\cos 45°\mathbf{i} + 5\sin 45°\mathbf{j} = 3.54\mathbf{i} + 3.54\mathbf{j}$.

4 $8\cos 50°\mathbf{i} + 8\sin 50°\mathbf{j} = 5.14\mathbf{i} + 6.13\mathbf{j}$.

5 **(a)** $8.06\ \text{m s}^{-1}$, 029.7°, **(b)** $7.81\ \text{m s}^{-1}$, 140.2°, **(c)** $12.0\ \text{m s}^{-1}$, 221.6°,
(d) $14.4\ \text{m s}^{-1}$, 303.7°.

6 **(a)** 8.54 m, **(b)** 159.4°.

7 $3.61\ \text{m s}^{-2}$, 303.7°.

EXERCISE 3C

1 $-0.3\mathbf{i} - 0.5\mathbf{j}$.

2 **(a)** $4.8\mathbf{i} + 2.4\mathbf{j}$, **(b)** $31.7\mathbf{i} + 8.6\mathbf{j}$.

3 $\mathbf{v} = (3 + t)\mathbf{i} + (-5 + t)\mathbf{j}$, $\mathbf{r} = \left(3t + \frac{t^2}{2}\right)\mathbf{i} + \left(-5t + \frac{t^2}{2}\right)\mathbf{j}$.

4 $36\mathbf{i} + 48\mathbf{j}$, $108\mathbf{i} + 144\mathbf{j}$.

5 $\mathbf{r} = 4t\mathbf{i} + (2 + 9t - 5t^2)\mathbf{j}$, $8\mathbf{i}$.

6 $r = (0.2 + \sqrt{3}t)\mathbf{i} + (0.1 + t)\mathbf{j}$, $1.76\mathbf{i} + 1\mathbf{j}$.

7 **(b)** 6.59 s, $r = 46.1\mathbf{i}$, **(c)** 54.1 m, **(d)** $33.3\ \text{m s}^{-1}$.

8 **(a)** 6 s, $120\mathbf{i}$, **(b)** 4 s.

9 **(a)** $\mathbf{v} = (4 - 0.02t)\mathbf{i} + (6 - 0.04t)\mathbf{j}$, $\mathbf{r} = (80 + 4t - 0.01t^2)\mathbf{i} +$
$(20 + 6t - 0.02t^2)\mathbf{j}$, **(b)** $416\mathbf{i} + 92\mathbf{j}$, **(c)** $0.4\mathbf{i} - 1.2\mathbf{j}$.

10 **(a)** $\mathbf{r}_H = 6t\mathbf{i} + (80 - 3t)\mathbf{j}$, $\mathbf{r}_A = (5t + 0.05t^2)\mathbf{i} + (10 + 0.025t^2)\mathbf{j}$,
(b) 20 s, $120\mathbf{i} + 20\mathbf{j}$.

11 $\mathbf{v} = (2 - 0.06t)\mathbf{i} + (3 - 0.04t)\mathbf{j}$, $\mathbf{r} = (40 + 2t - 0.03t^2)\mathbf{i} +$
$(20 + 3t - 0.02t^2)\mathbf{j}$, (b) 60 s.

12 **(a)** **(i)** 500 s, **(ii)** $10\,000\mathbf{i}$, **(iii)** $100.5\ \text{m s}^{-2}$.

4 Forces

EXERCISE 4B

1 **(a)** 5 N, 36.9°, **(b)** 14 N, 38.2°, **(c)** 12.8 N, 43.0°, **(d)** 4.35 N, 151.4°,
(e) 3.78 N, 84.9°.

2 **(a)** 90°, **(b)** 91.5°, **(c)** 152.3°, **(d)** 158.2°.

3 11 N, 1 N.

4 15.8 N.

5 11.1 N.

6 158.2°.

7 7.43 N, 53.8°.

EXERCISE 4C

1 **(a)** 1 N, **(b)** 5.96 N.

2 5.92 N.

3 $a = -7, b = -5$.

4 **(a)** $9\mathbf{i} + 17\mathbf{j}$, **(b)** 19.2 N, **(c)** 62.1°.

5 3.61 N, 123.7°.

6 $(\lambda - 4)\mathbf{i} + 4\mathbf{j}$, $\sqrt{\lambda^2 - 8\lambda + 32}$, $\lambda = 1$ or 7, 53.1°.

EXERCISE 4D

1 **(a)** 4.50, 5.36, **(b)** −7.25, 3.38, **(c)** −5.79, −6.89.

2 **(a)** $4.50\mathbf{i} + 5.36\mathbf{j}$, **(b)** $-7.25\mathbf{i} + 3.38\mathbf{j}$, **(c)** $-5.79\mathbf{i} - 6.89\mathbf{j}$.

3 **(a)** $-\frac{Mg}{2}, -\frac{Mg\sqrt{3}}{2}$, **(b)** $-W\sin\alpha, -W\cos\alpha$.

4 **(a)** 5.92 N, 88.5° below x, **(b)** 1.43 N, 82.3° below x,
(c) 7.21 N, 46.3° below x, **(d)** 4.56 N, 70.3° below x.

5 0 N.

6 **(a)** $-\frac{W}{\sqrt{2}}, -\frac{W}{\sqrt{2}}$, **(b)** $-\frac{W}{2}, -\frac{W\sqrt{3}}{2}$, **(c)** $-\frac{W\sqrt{3}}{2}, \frac{W}{2}$.

7 $-mg\sin\alpha$.

EXERCISE 4E

1 **(a)** $F_1 = 1.71$ N, $F_2 = 4.70$ N, **(b)** $F_1 = 4.01$ N, $F_2 = 17.7$ N,
(c) $F_1 = 5.22$ N, $F_2 = 7.04$ N.

2 **(a)** $F = 25$ N, $\theta = 73.7°$, **(b)** $F = 9.42$ N, $\theta = 15.8°$,
(c) $F = 8.64$ N, $\theta = 130.4°$.

3 **(a)** $F_1 = 7.83$ N, $F_2 = 6.21$ N, **(b)** $F_1 = 6.77$ N, $F_2 = 9.44$ N,
(c) $F_1 = 6.72$ N, $F_2 = 16.4$ N, **(d)** $F_1 = 10$ N, $F_2 = 3.34$ N,
(e) $F_1 = 4.79$ N, $F_2 = 6.16$ N, **(f)** $F_1 = 131$ N, $F_2 = 176$ N.

4 **(a)** $-8\mathbf{i} - 4\mathbf{j}$, **(b)** 8.94 N, 153.4° below **i**.

5 $a = -11, b = 0$.

6 **(a)** $\mathbf{F}_2 = -8\mathbf{i} - 9\mathbf{j}$, **(b)** $\mathbf{F}_3 = 2\mathbf{i} - 2\mathbf{j}$.

7 None.

8 24.5 N, 42.4 N.

9 38.7 N, 72.7 N.

10 5.57 N, 17.9°.

11 $P = 8.39$ N, $R = 13.1$ N.

12 11.2 N, 26.6°.

13 124.2°, 97.2°, 138.6°.

14 1.81 kg.

15 27.1 N, 24.0 N.

16 $-90\mathbf{i} + 40\mathbf{j} + 132\mathbf{k}$, 165 N.

17 $-88\mathbf{i} - 128\mathbf{j}$, 155 N.

EXERCISE 4F

1 0.255.

2 **(a)** No motion, **(b)** slides, **(c)** no motion.

3 0.567.

4 73.6 N.

5 51.8 N, 185°, 0.529.

6 0.5.

7 88.2 N.

8 **(a)** 0.085, **(b)** 30.8 N.

EXERCISE 4G

1 **(a)** Slides, **(b)** no motion, **(c)** slides.

2 **(a)** 7.37 N, **(b)** 24.3 N, **(c)** 44.2 N.

3 **(a)** 26.7 N, **(b)** 42.7 N, **(c)** 70.4 N.

4 34.6 N.

5 **(a)** 28.3 N, **(b)** 181 N, **(c)** 671 N.

6 **(a)** 29.2 N, **(b)** 7.88 N, **(c)** 14.5 N.

7 $F = 21.9$ N, $R = 77.9$ N, 0.281.

8 **(a)** $R = 53.9$ N, $F = 33.6$ N, $\mu \geq 0.623$,
(b) $R = 36.8$ N, $F = 6.59$ N, $\mu \geq 0.179$,
(c) $R = 36.8$ N, $F = 38.4$ N, $\mu \geq 1.04$.

5 Newton's laws of motion

EXERCISE 5A

1 **(a)** $X = 6$ N, $Y = 5$ N, **(b)** $X = 10$ N, $Y = 12$ N, **(c)** $X = 69.3$ N, $Y = 69.3$ N.

2 **(a)** 10 N, 36.9°, **(b)** 15.7 N, 60°, **(c)** 8.66 N, 60°.

3 1350 N.

4 0.245.

5 **(a)** $T = 115$ N, **(b)** $R = 112$ N.

6 210 N.

7 49.5 m s^{-1}.

8 17.9 m s^{-1}.

9 43.39 m s^{-1}.

EXERCISE 5B

1 0.5 m s^{-2}.

2 6 N.

3 $(2.5\mathbf{i} + \mathbf{j})$ m s^{-2}.

4 0.7 m s^{-2}.

5 **(a)** 2970 N, **(b)** 2880 N, **(c)** 2910 N.

6 **(a)** $a = 8$ m s^{-2}, **(b)** $a = 0.8$ m s^{-2}, $R = 98$ N, **(c)** $T = 47.3$ N, $R = 74.3$ N, **(d)** 103 N, **(e)** $\theta = 30°$, $a = 0.329$ m s^{-2}, **(f)** $T = 66.5$ N, $a = 3$ m s^{-2}.

7 4.77 m s^{-2}.

8 87.1°.

9 4167 N.

10 4.9 m s^{-2}.

11 4 s.

12 33.3 N, 0.195.

13 11.8°.

14 61 500 N, 49 000 N, 24 000 N, 10.3 m.

15 $P = 10$ N.

16 $(13\mathbf{i} + 20\mathbf{j})$ m.

17 **(a)** 12.1°, **(b)** 116 000 N, 0.0058 m s^{-2}.

18 **(a)** 253 N, **(b)** 10.2 N, **(c)** 0.340 m s^{-2}.

19 **(a)** 8 m, 7.5 s, **(b)** first stage 4240 N.

20 **(b)** 2428 N, **(c)** will stop accelerating.

21 **(a)** $\frac{m(a+g\sqrt{3})}{2}$, **(b)** 17.0 m s^{-2}, 5.66 m s^{-2}, unlikely but possible, **(c)** no.

22 **(a)** 16 000 m, **(b)** $233\frac{1}{3}$ s, **(c)** 10 000 N in opposite direction to motion.

23 **(a)** 194 N, **(b)** 3.24 m s^{-2}.

EXERCISE 5C

4 $T = 2863$ N, $R = 803$ N, $W = 1960$ N.

6 Kinematics and variable acceleration

EXERCISE 6A

1 **(a)** $83\frac{1}{3}$ m, **(b)** $v = \frac{t^2}{2} - \frac{t^3}{30}$, **(c)** $a = t - \frac{t^2}{10}$,
(d) Increases from 0 initially, to a maximum at $t = 5$ and then decreases to 0 when $t = 10$.

2 **(a)** $v = 2t - \frac{t^2}{20}$, $a = 2 - \frac{t}{10}$, **(b)** $t = 20$,
(c) 20 m s^{-1}, **(d)** 267 m.

3 **(a)** Lift comes to rest,
(b)

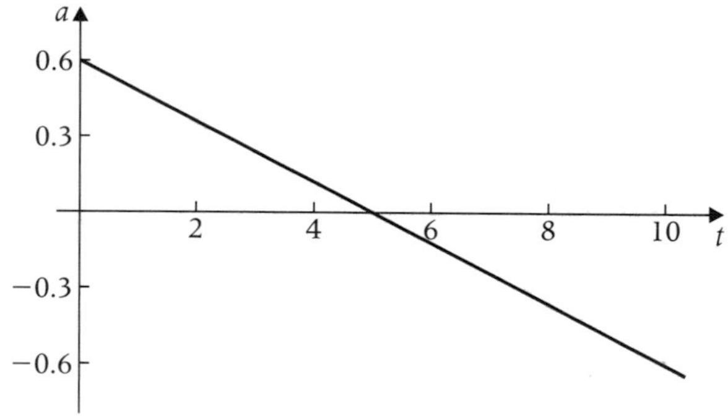

4 **(a)** $0 \leqslant t \leqslant 15$, **(b)** 112.5 m,
(c)

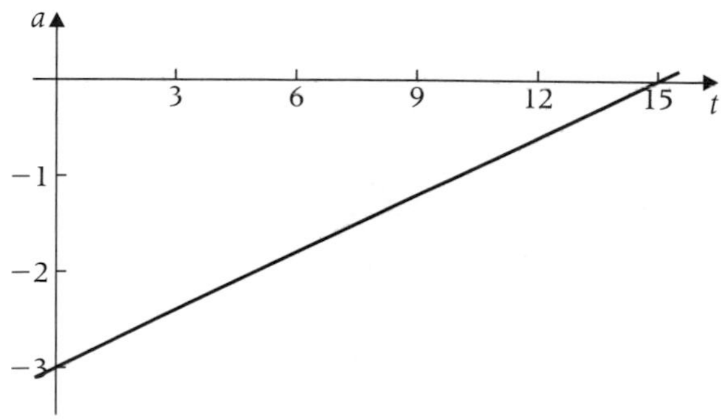

5 243 m.

6 **(a)** $v = 0.2 \cos(0.5t)$, **(b)** 0.2 m s^{-1},
(c) -0.0841 m s^{-2}, **(d)** $-0.1 \leqslant a \leqslant 0.1 \text{ m s}^{-2}$.

7 **(a)** 0, **(b)** $0 \leqslant t \leqslant 800\pi$,
(c) 300 m, **(d)** $0.000\,234 \text{ m s}^{-2}$.

8 **(a)** $A = 15$, **(b)** $k = \frac{4}{3}$,
(c)

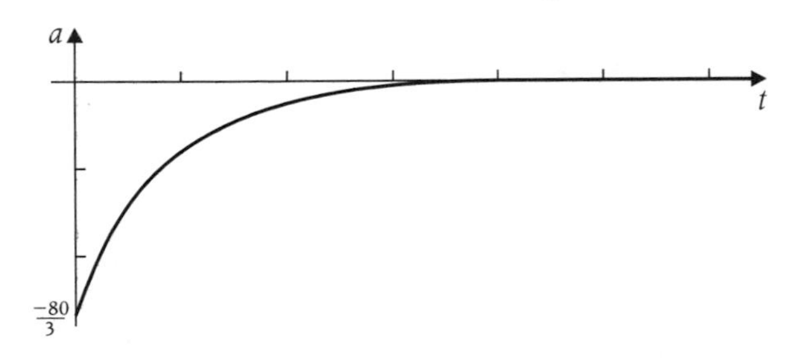

9 **(a)** 4 m, **(b)** 6 m s^{-1}.

10 **(a)** 0 m s^{-1}, 40 m s^{-1},
(b)

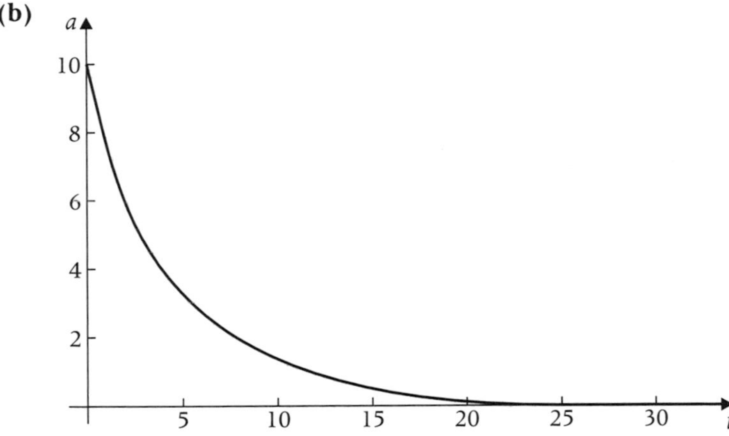

11 **(b)** $-(kU + g)$

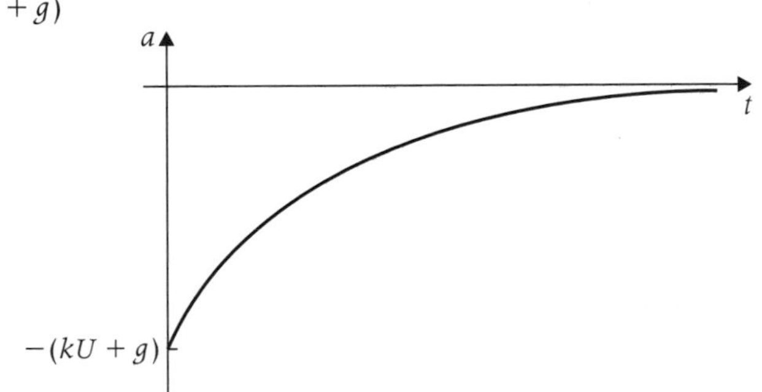

12 **(a)** $v = 5t - \frac{t^3}{5}$, $a = 5 - \frac{3t^2}{5}$, **(b)** $0 \leqslant t \leqslant 5$, 31.25 m,
(c) decreases to -10, 9.62 m s^{-1},
(d) falls from rest, no air resistance.

EXERCISE 6B

1 **(a)** $v = \frac{t^3}{300}$, **(b)** $\frac{5}{12}$ m s^{-1}, **(c)** $\frac{25}{48}$ m.

2 20 m s^{-1}, 267 m. **3** 38.3 m. **4** 11.1 m s^{-1}, 42.1 m.

5 **(b)** 17.5 m s^{-1}, **(c)** 183 m.

6 **(a)** $v = 2t - \frac{t^2}{20}$, **(b)** $t = 20$ s, $v = 20$ m s^{-1},

(c) Speeds up for first 20 s and slows down for last 10 s,

(d) 450 m.

7 **(a)** $a = 1.8 \cos(3t)$, **(b)** $x = 1 - 0.2 \cos(3t)$,

(c)

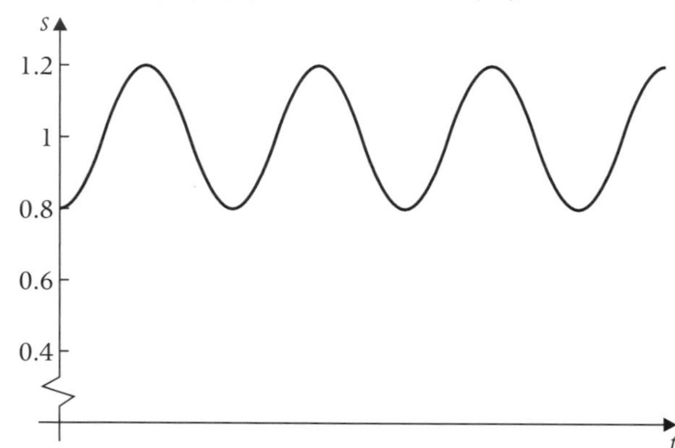

8 **(a)** $v = 4.9(1 - e^{-2t})$, $s = 2.45e^{-2t} + 4.9t - 2.45$,

(b)

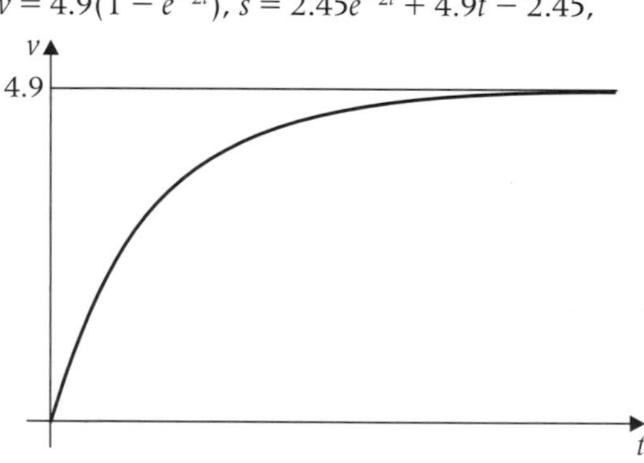

(c) 95.6 m.

9 18.7 m.

10 **(a)** $v = \frac{2}{5\pi} \sin(100\pi t)$, **(b)** $\frac{2}{5\pi} = 0.127$ m s^{-1},

(c) 0.000 405 m.

11 **(a)** $x = \frac{3}{2} - \frac{3}{2} \cos(2t) - \sin(2t)$, **(b)** $\sqrt{13}$ m s^{-1}.

EXERCISE 6C

1 **(a)** $\mathbf{v} = 2t\mathbf{i} + (12t - 1)\mathbf{j}$, $\mathbf{a} = 2\mathbf{i} + 12\mathbf{j}$,

(b) $\mathbf{r} = 11\mathbf{i} + 96\mathbf{j}$, $\mathbf{v} = 8\mathbf{i} + 47\mathbf{j}$.

2 **(a)** $\mathbf{v}_A = 30\mathbf{i} + (6t - 120)\mathbf{j}$, $\mathbf{v}_B = 20\mathbf{i} + 40\mathbf{j}$,

(b) $v_B = 44.7$ m s^{-1}, **(c)** $t = 30$, 756 m.

3 $\mathbf{v} = 2t\mathbf{i} + 2\mathbf{j}$, $\mathbf{a} = 2\mathbf{i}$.

4 **(a)** $\mathbf{v} = \left(4 - \frac{2t}{5}\right)\mathbf{i} + 10\mathbf{j}$, **(b)** $t = 20$, 200 m,
(c) $t = 10$, 10 m s^{-1}, **(d)** $\frac{2}{5}$ m s^{-1} due west.

5 **(a)** $\mathbf{v} = (2t - 8)\mathbf{i} + (6t^2 - 10t + 6)\mathbf{j}$,
(b) $\mathbf{F} = 6\mathbf{i} + (36t - 30)\mathbf{j}$,
(c) $t = 4$, $-14\mathbf{i} + 72\mathbf{j}$.

6 **(a)** $t = 2$, $v = 20.2$ m s^{-1}, **(b)** $t = 5$, $\mathbf{a} = -\mathbf{j}$.

7 $38.3\mathbf{i} + 16.7\mathbf{j}$.

8 **(a)** 3.2 kg, **(b)** $28.3\mathbf{i} + 46.9\mathbf{j}$.

9 $\mathbf{v} = (2 + t)\mathbf{i} + (5 + 0.1t^2)\mathbf{j}$, $\mathbf{r} = \left(2t + \frac{t^2}{2}\right)\mathbf{i} + \left(5t + \frac{t^3}{30}\right)\mathbf{j}$.

10 **(a)** $\mathbf{v} = (3 + t^2)\mathbf{i} + \left(6 - \frac{5t^2}{2}\right)\mathbf{j}$, **(b)** $\mathbf{r} = \left(3t + \frac{t^3}{3}\right)\mathbf{i} + \left(6t - \frac{5t^3}{6}\right)\mathbf{j}$.

11 **(a)** $\mathbf{v} = 4t\mathbf{i} - \frac{t^2}{2}\mathbf{j}$, **(b)** $\mathbf{r} = (5 + 2t^2)\mathbf{i} - \left(10 + \frac{t^3}{6}\right)\mathbf{j}$.

12 32 m s^{-1}, 256 m s^{-2}.

13 **(a)** $-4\mathbf{i} + 0\mathbf{j}$, $0\mathbf{i} - 6\mathbf{j}$, $16\mathbf{i} + 0\mathbf{j}$, **(b)** 16 m s^{-2}, 12 m s^{-2}.

14 **(a)** 3 m s^{-2} upwards,
(b) **(i)** $4\mathbf{i} + 6\mathbf{j} + 6\mathbf{k}$, **(ii)** 9.38 m s^{-2},
(c) 39.8° above horizontal.

15 **(a)** 20 m s^{-1}, 20 m s^{-1}, 20 m s^{-1}, speed is constant,
(b) $100\mathbf{i} + 200\mathbf{j} + 0\mathbf{k}$, $100\mathbf{i} + 210\mathbf{j} + 0\mathbf{k}$, 10.5 m.

16 $t = 0$, $t = \pi$.

17 **(a)**

	A	B
	$0\mathbf{i} + 0\mathbf{j}$, $20\mathbf{i} - 1.6\mathbf{j}$	$0\mathbf{i} + 0\mathbf{j}$, $20\mathbf{i} - 0\mathbf{j}$
	$100\mathbf{i} - 4\mathbf{j}$, $20\mathbf{i} + 0\mathbf{j}$	$100\mathbf{i} - 4\mathbf{j}$, $20\mathbf{i} + 0\mathbf{j}$
	$200\mathbf{i} + 0\mathbf{j}$, $20\mathbf{i} + 1.6\mathbf{j}$	$200\mathbf{i} + 0\mathbf{j}$, $20\mathbf{i} + 0\mathbf{j}$

(b) Model **B** is better as the car will not leave the road when overtaking is complete or start and finish velocities are parallel to the road.

18 **(a)** $\mathbf{v} = 25\mathbf{i} + 5\sin\left(\frac{\pi t}{10}\right)\mathbf{j}$, **(b)** $v = 5\sqrt{25 + \sin^2\left(\frac{\pi t}{10}\right)}$,
(c) Max speed when $t = 5$, 15. Min speed when $t = 0$, 10, 20.
Speed increases and decreases while going up.
Speed increases and decreases while going down.
(d) Would expect speed to decrease on way up and increase on way down.

7 Energy

EXERCISE 7A

1 4.8 J.

2 3.75×10^{10} J.

3 3.6 J, 14.4 J.

4 4 725 000 J.

5 **(a)** 28 m s^{-1}, **(b)** 19.6 J.

6 **(a)** 7.68 m s^{-1}, **(b)** 1750 J.

EXERCISE 7B

1 **(a)** 320 000 J,
(b) **(i)** 320 000 J, 25.3 m s^{-1},
(ii) 324 500 J, 25.5 m s^{-1}.

2 **(a)** 50 J, **(b)** 50 J, **(c)** 16.7 N.

3 **(a)** 62.72 J, 7.92 m s^{-1}, **(b)** 30.72 J, 5.54 m s^{-1}.

4 **(a)** 25 000 J, **(b)** 16 J, **(c)** 24 984 J, **(d)** 499.68 N.

5 0.8575 J, 56.25 cm.

6 **(a)** 1 340 000 J, 46.3 m s^{-1}, **(b)** 777 500 J, 1555 N.

7 **(a)** 300 000 J, **(b)** 235 200 J, **(c)** 64 800 J, **(d)** 20 m,
(e) air resistance is not constant, usually some function of speed: the effect would be a greater stopping distance.

8 **(a)** 432 J, 108 N, **(b)** 1548 N.

9 **(a)** 249.9 N, **(b)** 1.45 cm.

10 48 N, 98.0 m s^{-1}.

EXERCISE 7C

1 **(a)** 2450 J, **(b)** 9.90 m s^{-1}.

2 **(a)** 54 J, **(b)** 6.89 m.

3 **(a)** 416 J, **(b)** 440 J, 17.1 m s^{-1}, **(c)** 15.1 m s^{-1}.

4 **(a)** 630 J, **(b)** 750 J, 5.00 m s^{-1}, **(c)** 1.28 m.

5 **(a)** 21.6 J, **(b)** 19.2 J, **(c)** 6.53 m.

6 **(a)** 32 400 J, **(b)** 20 640 J, **(c)** 12.7 m s^{-1}.

7 286 J, 6.90 m s^{-1}.

8 **(a)** 37 800 J, **(b)** 1890 N, **(c)** 13.2 m.

9 98 700 J, 13.4 m s^{-1}.

10 **(a)** 2695 J, 9.90 m s^{-1}, **(b)** 9.46 m s^{-1}, **(c)** 80.5°.

11 **(a)** 1728 J, 6.57 m s^{-1}, no air resistance,
(b) 2.20 m, **(c)** perpendicular to direction of motion.

12 **(a)** 431 N, **(b)** 25.7 m,
(c) use a variable resistance force, model the roller coaster as a number of connected particles.

EXERCISE 7D

1 **(a)** 0.392 m, **(b)** 0.235 m, **(c)** 0.0392 m.

2 0.5 m.

3 156.8 N .

4 68.0 grams.

5 **(a)** 0.392 cm, **(b)** 0.196 cm, **(c)** 0.784 cm.

EXERCISE 7E

1 **(a)** 0.8 J, **(b)** 2.4 J.

2 7.07 m s^{-1}.

3 2.21 m s^{-1}.

5 **(a)** 1.28 m s^{-1}, **(b)** 16.7 cm.

6 40.5 cm, 2.21 m s^{-1}.

7 $\lambda = \frac{3mg}{2}$.

8 **(a)** $v = \sqrt{\frac{800 - 70\,875x^2}{162}}$, **(b)** 10.6 cm.

9 **(a)** 0.196 cm, **(b)** 11.2 cm.

10 50.8 cm.

11 0.657 m s^{-1}.

12 3.32 m s^{-1}, 0.963 m.

13 7.07 m s^{-1}.

14 $\sqrt{2ga}$, $\sqrt{6ga}$.

15 **(a)** $9\sqrt{\frac{ga}{80}}$, **(b)** $\frac{5a}{4}$.

EXERCISE 7F

1 669 N.

2 1.64 m s^{-2}.

3 102 km.

4 **(a)** 785 N, **(b)** almost identical.

5 **(a)** 3.67×10^{24} N, **(b)** 3.67×10^{24} N.

EXERCISE 7G

1 3.08×10^{10} J.

2 **(a)** 7.38×10^{6} J, **(b)** yes, as values almost the same.

3 2.51×10^{9} J.

4 2.32×10^{6} J.

5 1.25×10^{11} J.

EXERCISE 7H

1 **(a)** 94 080 J, **(b)** 784 W.

2 100.8 W.

3 9×10^{5} W.

4 68.1 W.

5 **(a)** 900 N, **(b)** 675 N, **(c)** 0.525 m s^{-2}.

6 **(a)** $\frac{75v}{4}$, **(b)** 11 700 W, **(c)** 27.1 m s^{-1}.

7 31.2 m s^{-1}, 80.2 m s^{-1}.

8 **(a)** $\frac{400v}{27}$, **(b)** 37.2 m s^{-1}, **(c)** 54.5 m s^{-1}.

9 **(a)** 34.6 m s^{-1}, **(b)** 24.5 m s^{-1}.

10 **(a)** 200 N, **(b)** 427 000 J.

11 **(b)** 1.33 m s^{-1},
(c) **(i)** motion very slow, so little air resistance,
(ii) would accelerate for ever without air resistance.

12 **(a)** $30v$ N, **(b)** 19%,
(c) small reduction leads to larger fuel savings, **(d)** 28.6 m s^{-1}.

13 **(a)** 625 N, **(b)** 32.7 m s^{-1}.

14 7500 W, 5 m s^{-2}.

15 (a)

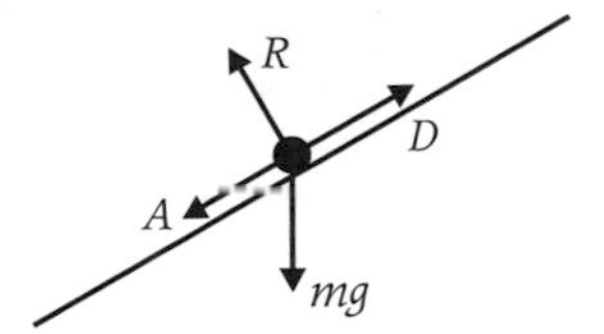

There are really 4 reaction forces.

(c) 50 000 W.

8 Circular motion

EXERCISE 8A

1 (a) $\frac{1}{60}$ rpm, **(b)** $\frac{\pi}{1800}$ rad s^{-1}.

2 $\frac{\pi}{3}$ rad s^{-1}.

3 51 600 m s^{-1}.

4 2990 m s^{-1}.

5 52.4 m s^{-1}, 13 700 m s^{-2}.

6 35.5 N.

7 2.5 m.

8 1.54 m s^{-2}.

9 20.9 m s^{-1}, 2190 m s^{-2}.

10 (a) $\frac{\pi}{3}$ rad s^{-1}, **(b)** $\frac{2\pi}{3}$ m s^{-1}, $\frac{\pi}{2}$ m s^{-1}.

EXERCISE 8B

1 1750 N.

2 3.6 N towards the centre of the circle.

3 72 N.

4 11.6 m s^{-1}.

5 2 N.

6 576 N.

7 $2\pi\sqrt{\frac{a^3}{k}}$.

8 38.1 N.

9 3290 N, 0.336.

10 6.26 rad s^{-1}.

11 **(a)** Coin slides, **(b)** 4.4 cm, **(c)** no change.

12 0.227, 1.19m N.

13, **(b)** 6 m s^{-1}, **(c)** 3 rad s^{-1}, **(d)** 72 N.

14 0.340.

15 18.8 m s^{-1}.

16 1.98 rad s^{-1}.

17 **(a)** 49.3 m s^{-2}, 25.2 N, **(b)** 0.639 s.

18 **(a)** 125 m s^{-2}, **(b)** path is a tangent to the end of the strip.

19 0.33.

20 **(a)** 6000 N, **(b)** $\mu \geqslant \frac{25}{49}$, **(c)** 21.7 m s^{-1}, **(d)** air resistance.

21 **(a)**

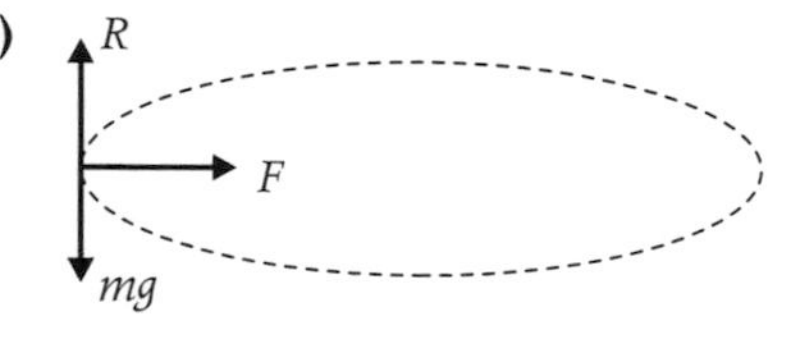

(b) 16.3 m, **(c)**

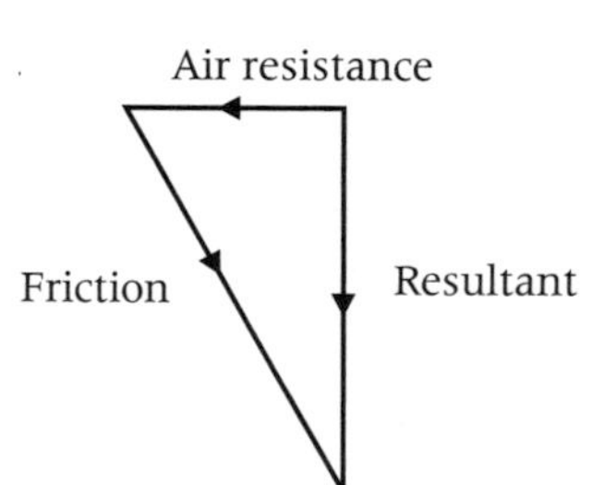

EXERCISE 8C

1 **(a)** 20.9 N, 4.57 rad s^{-1}, **(b)** 50 N, 66.9°, **(c)** 0.272 m, 39.2 N.

2 20.0 N, 4.47 rad s^{-1}.

3 82.9 N, 65.5 cm.

4 **(a)** 48 N, **(b)** 35.2°, **(c)** 1.73 m.

5 56.6 N, 1.68 m s^{-1}.

6 2.56 rad s^{-1}.

7 0.959 rad s^{-1}.

8 $\sqrt{\frac{g\sqrt{3}}{r}}$.

9 $\frac{g}{l\omega^2}$.

10 $\frac{g}{2\pi^2}$.

11 $v = \frac{\sqrt{15ag}}{2}$, $R = 4mg$.

14 **(a)** 10 000 N, **(b)** $\alpha = \tan^{-1}\left(\frac{v^2}{rg}\right)$, $L = m\sqrt{\frac{v^4}{r^2} + g^2}$,
(c) reduce r then α increases and L increases.

15 **(a)** **(i)** 453 N, **(ii)** 5.66 m s^{-2}, **(iii)** 3.36 m s^{-1},
(b) 0.553 s, due to the actual size of the child,
(c) 2.73 m, between 2 and 2.73 m from O.

16 **(a)** Light, inextensible, **(c)** 86.6 N, 52.0 N.

17 **(b)** $R = m\left(\frac{g}{\sqrt{2}} - \frac{v^2}{l}\right)$, **(c)** particle leaves contact with the cone.

18 **(a)** 19.7 m s^{-2},
(b) R = 1380 N, F = 686 N.

19 **(a)** $\tan\alpha = 5$,
(c) $3.83 \leqslant v \leqslant 4.43$ m s^{-1},
(d) 0.0735 J,
(e) rotation of the coin.

EXERCISE 8D

1 34.2 m s^{-1}, 26.7°.

2 13.9 m s^{-1}.

4 25.0 m s^{-1}.

5 22.7 cm.

6 2770 N, 10 900 N, 0.254.

7 0.4, 21.8°, 35.2 m s^{-1}.

8 **(a)** 9600 N,
(b) 9180 N.

EXERCISE 8E

1 **(a)** 3192 N, **(b)** 3198 N, **(c)** 3204 N.

2 910 N.

3 **(a)** 2020 N,
(b) when disc has rotated 57.8° from lowest position.

4 A 54 000 N. B 33 200 N. C 72 900 N. D 68 700 N.

5 2.42 m s^{-1}.

Exam style practice paper

1 **(a)** $1\ \text{m s}^{-2}$, **(b)** $20\ \text{m s}^{-1}$, **(c)** 2.7 N.

2 **(a)** $\frac{1}{20}\mathbf{i} + \frac{1}{30}\mathbf{j}$, **(b)** 433 m, **(c)** greater than 433 m.

3 **(a)** $2.5\ \text{m s}^{-2}$, **(c)** $14.3°$,

(d) no change, **(e)** $-0.514\ \text{m s}^{-2}$.

4 **(a)** 705 m, **(b)** $\mathbf{v} = 50\mathbf{i} - t\mathbf{j} + 1.6t\mathbf{k}$,

(c) 12.5 s, **(d)** 50 s.

5 **(a)** 1000 J, **(b)** $1.98\ \text{m s}^{-1}$, **(c)** 4.33 m.

6 **(b)** $26\ \text{m s}^{-1}$, $9\ \text{m s}^{-1}$,

(c) First model as the top speed for the other is very low.

7 **(a)** $\sqrt{2}\,mg$, **(b)** $v \geqslant \sqrt{rg}$, **(c)** $v = \sqrt{rg(1 + \sqrt{2})}$.

8 **(a)** $v = e^{-0.2t}(100 \cos 20t - \sin 20t)$, **(b)** $96.9\ \text{m s}^{-1}$.

Index

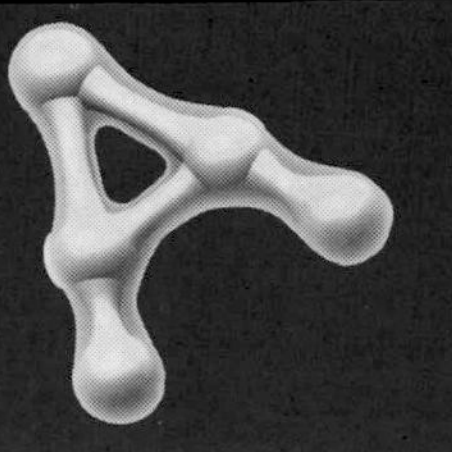